全国高等职业教育专业英语系列规划教材

计算机英语

JISUANJI YINGYU

主　编　姚姗姗　陶晋宇
副主编　毛琰虹
参　编　鲁宵昳　杨　阳　袁晓冬

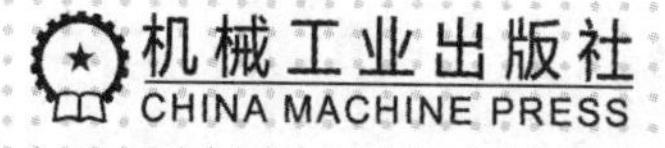

本书依据计算机行业典型工作过程来组织内容，可供计算机、网络通信类各专业学生使用，旨在培养学生使用英语处理计算机相关业务的能力。本书采用双色印刷，共12个单元，涵盖建立业务关系、产品设计、产品测试、产品销售、电子商务、计算机安全、故障与排除、售后服务、行业展望等内容，可满足计算机行业从业人员工作过程中基本的英语交际需求。

本书每单元均由四部分组成，包括：学习目标、行业对话、行业阅读、行业写作。学习目标列出本单元所涉及的主要内容与技能要求，帮助学生在学习之前了解本单元的基本要求。行业对话一般涵盖5～6段对话，强调语言的输入与输出，帮助学生熟悉本工作过程典型的交际场景，从而熟练掌握与该主题相关的语言表达方式。行业阅读由三篇文章组成，前两篇为实用阅读，根据单元主题提供相应的阅读材料，并配有词汇表、难句注释、课后习题及课文翻译，培养学生对于行业阅读材料的理解、翻译能力。第三篇课文作为补充性阅读材料，供学生在课下自主学习使用。行业写作根据单元主要内容，选取该工作流程中所涉及的主要实用写作文体，提供应用文范例，配以练习，旨在培养学生参照范例使用英语书写特定问题的能力。

本书从计算机行业从业人员实际工作需要出发进行设计和编写，选材富有时代性、内容丰富、点面结合，旨在提高工作人员在本领域的涉外业务交际能力，提高学习者的就业能力和职业竞争力。

凡选用本书作为教材的教师，均可登录机械工业出版社教育服务网 www.cmpedu.com 下载本教材配套电子课件，或发送电子邮件至 cmpgaozhi@sina.com 索取。咨询电话：010-88379375

图书在版编目（CIP）数据

计算机英语 / 姚姗姗，陶晋宇主编. —北京：机械工业出版社，2017.5

全国高等职业教育专业英语系列规划教材

ISBN 978-7-111-56893-3

Ⅰ.①计… Ⅱ.①姚… ②陶… Ⅲ.①电子计算机-英语-高等职业教育-教材 Ⅳ.①TP3

中国版本图书馆 CIP 数据核字（2017）第108990号

机械工业出版社（北京市百万庄大街22号 邮政编码100037）

策划编辑：赵志鹏　　责任编辑：赵志鹏　杨　洋

封面设计：马精明　　责任印制：常天培

责任校对：赵志鹏

保定市中画美凯印刷有限公司印刷

2017年4月第1版第1次印刷

184mm×260mm · 11印张 · 237千字

0001-3000册

标准书号：ISBN 978-7-111-56893-3

定价：29.80元

凡购本书，如有缺页、倒页、脱页，由本社发行部调换

电话服务

服务咨询热线：010-88379833

读者购书热线：010-88379649

网络服务

机 工 官 网：www.cmpbook.com

机 工 官 博：weibo.com/cmp1952

教育服务网：www.cmpedu.com

金 书 网：www.golden-book.com

前 言

新时代信息技术的专业人才，不仅需要具备扎实的专业知识与技能，而且需要掌握一定的行业英语知识，用以获取本专业的前沿知识来拓宽专业知识面。本书以“工学结合、能力为本”的职业教育理念为指导，将语言学习与职业技能培养有机结合，在帮助学生打好语言基础的同时，注重培养学生在不同职业场景中的英语交际能力，确保课程内容真正体现职业性和实用性。

本书供计算机和网络通信类各专业在完成基础英语阶段学习后使用，可基本满足计算机行业从业人员工作过程中的英语交际需要，并为学生进一步学习专业英语打好基础。本书具有以下特点：

采用“英语教师+专业教师+行业人员”的开发模式，充分发挥各方专长。

选用计算机行业中真实的语言材料作为学习材料，以计算机行业的工作过程、典型工作环节和场景为参照来组织内容，根据主要工作任务所需的英语知识和技能，设计英语学习任务，注重时代性、信息性与实用性。注重培养学生用英语处理与计算机相关业务的能力，兼顾交际技能、职业技能和自主学习能力的培养。

本书采用双色印刷，共12个单元，供1学期使用。在每单元开始前有学习目标，以便对学生学习本单元知识进行引导；其后设以情景对话；每课配有相应的词汇表、注释、习题、扩展阅读、参考译文及习题答案，可供学生检查学习效果与自测使用。在内容选择和编排上，本书充分考虑了当前计算机专业英语发展的现状以及高职院校的实际需求，遵循了由浅入深、循序渐进的原则。选材力求紧跟计算机行业的发展步伐，做到内容新、知识面广、词汇量大。全书内容通俗易懂、内容丰富、结构新颖、重点突出。

本书可作为高等院校计算机及相关专业的行业英语教材，也可作为广大科技人员学习计算机专业英语知识或参加有关计算机专业英语知识考试的参考用书。

本书由河南工业职业技术学院一线优秀教师编写。具体分工如下：姚姗姗负责全书的体系结构，并编写第8、11、12单元及其习题、译文和答案；陶晋宇编写第1、2、3单元及其习题、译文和答案；毛琰虹编写第4、5、6单元及其习题、译文和答案；杨阳编写第7单元及其习题、译文和答案；鲁宵昳编写第9单元及其习题、译文和答案；袁晓冬编写第10单元及其习题、译文和答案。全书由陶晋宇统稿。

由于时间仓促，且编者水平有限，加之计算机行业发展迅速，书中难免有疏漏和不足之处。我们衷心地希望得到广大读者的批评指正，以使本书在教学实践中进一步完善。

编　者

目 录

Unit 1 The Overview of Computers

Learning Objectives

In this unit you should get familiar with:

- The development of computers;
- The computer system;
- Understanding and filling in registration forms.

Speaking

Dialogue 1

A: Where did the web come from?

B: It started in 1989 at a Laboratory in Europe known as CERN where physicists around the world work together.

A: Why is it so popular?

B: Because it is easy to use and connects people around the world who want to locate information and share knowledge.

A: Thanks. I think I'll go surf for a new salad recipe.

Dialogue 2

A: How do you think of the network?

B: One of the simplest networks involves two computers and one printer.

A: Really? It's so supernatural.

B: Do you know what the advantage of network is?

A: Yes, sharing resources between computers.

B: That's right.

Dialogue 3

A: What is World Wide Web?

B: That's just like saying the Internet.

A：OK. That's not so difficult to understand.

B：Do you know how to log into Internet?

A：Yes，of course. It's very easy.

B：Can you teach me?

A：Enter your address and password，it's OK.

Dialogue 4

A：Hi，Mike.

B：Hi，what can I do for you?

A：Do you know how useful the computer is in our life?

B：Of course，I do know about it. My major is Computer Science.

A：Let's discuss about it.

B：Computer helps us edit our documents. You can modify your document freely. It can copy and paste the useful words，and save your time.

Dialogue 5

A：Is this your school computer room? Who are these people?

B：They are students of the Russian Department.

A：What are they doing?

B：They are learning to use computers.

A：Do you often work here?

B：Yes，we work here once a week. We are all keen to learn modern techniques.

Dialogue 6

A：Would you please write your name on this list?

B：Then what?

A：I will call you when a computer is free.

B：How do I log on to the computer?

A：Use the number on the back of your library card.

B：Thanks. I'll be sitting over there.

Text Reading

Text A

The Computer's Development

In the history of computers，there are a few development stages. Therefore，several

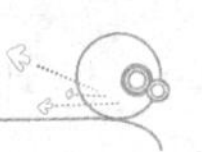

computer generations occur in the history.

The first electronic digital computer was born in America in 1946 and its basic elements were vacuum tubes.[1] Through the 1950s, several others were built. They were the first generation of computers, huge, heavy, expensive and slow, as well as using much more power than today's, but they still made great contributions to computer science, such as the concepts of stored programs, random access.[2] They made a basic model of modern electronic computers.

The invention of transistors not only produced small portableradios, but also bore the second generation of computers. They became small, light, less expensive, but they were not yet small and cheap enough to enter families.

In 1960s, integrated circuits came. Integrated circuits meant that huge complicated circuits and millions of their elements were only made on a small semiconductor chip.[3] They were introduced into the third generation of computers. Their typical models were the system 360line of IMB computers. Special, large scale integrated circuits made digital computers so popular that most middle class families could easily afford them. It is why you can see PCs everywhere.

Computers have already changed our life in many ways. The unprecedented access to information provided by computers has changed society's privacy landscape.[4] Credit cards, banks and telephone companies record users' business activities. Internet firms provide free services such as online searches, maps, and e-mail save information typed in by users and sort it by unique identification numbers in the machines.

With the development of science and technology, biological computers and quantum computers will emerge out in near future. New generations of computers will be born.

Notes:

[1] The first electronic digital computer was born in America in 1946 and its basic elements were vacuum tubes.

第一台数字电子计算机 1946 年诞生于美国，它的基本元件是电子管。

1946 年：ENIAC（Electronic Numerical Integrator and Computer）诞生，这是第一台真正意义上的数字电子计算机。开始研制于 1943 年，完成于 1946 年，负责人是 John W. Mauchly 和 J. Presper Eckert，重 30 吨，用了 18 000 个电子管，功率 25 kW，主要用于计算弹道和氢弹的研制。

[2] They were the first generation of computers, huge, heavy, expensive and slow, as well as using much more power than today's, but they still made great contributions to computer science, such as the concepts of stored programs, random access.

它们是第一代计算机，体积大、重量重、价格贵、速度慢且消耗的能量比现在计算机消耗的能量要多得多，但它们依然为计算机科学做出了重大的贡献，如程序存储和随机访问等。

make great contributions to 意为“对……做出巨大贡献”。

[3] In 1960s, integrated circuits came. Integrated circuits meant that huge complicated circuits and millions of their elements were only made on a small semiconductor chip.

20 世纪 60 年代，集成电路问世。集成电路意味着巨大的复杂电路和上百万个元件可以做在一小片半导体的芯片上。

[4] The unprecedented access to information provided by computers has changed society's privacy landscape.

计算机为人们提供了史无前例的获取信息的渠道，改变了社会的隐私状况。

access to 意思为“进入……的入口，通路”。

Vocabulary

generation	[dʒenəˈreɪʃ(ə)n]	*n*.	一代人；代（约 30 年），时代
occur	[əˈkɜː]	*v*.	发生；出现；闪现
electronic	[ɪˌlekˈtrɒnɪk]	*adj*.	电子的；电子操纵的
digital	[ˈdɪdʒɪt(ə)l]	*adj*.	数字的；数据的
vacuum	[ˈvækjʊəm]	*n*.	真空，空白
tube	[tjuːb]	*n*.	管，管状物；电子管
huge	[hjudʒ]	*adj*.	巨大的；庞大的
contribution	[kɒntrɪˈbjuːʃ(ə)n]	*n*.	贡献，捐赠，捐助
concept	[kɒnsept]	*n*.	观念，概念；观点；思想
random	[ˈrændəm]	*adj*.	任意的；随机的；胡乱的
invention	[ɪnˈvenʃən]	*n*.	发明；发明物
transistor	[trænˈzɪstə]	*n*.	晶体管；半导体收音机
portable	[ˈpɔrtəbl]	*adj*.	手提的；轻便的
integrate	[ˈɪntɪgreɪt]	*v*.	使一体化；使整合
circuit	[ˈsɜːkɪt]	*n*	电路，线路
complicated	[ˈkɒmplɪkeɪtɪd]	*adj*.	结构复杂的；混乱的

Exercises

Ⅰ. **Answer the following questions briefly according to the text.**

1. When was the first electronic digital computer launched?
2. What functions did the first electronic computer have?
3. What causes the production of the second generation of computers?
4. What's the meaning of integrated circuits?
5. How does the computer change our life?

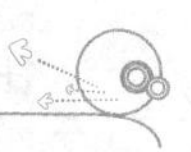

Ⅱ. Fill in the table below by giving the corresponding Chinese or English equivalents.

electronic digital computer	
	电子管
computer science	
	晶体管
integrated circuits	
	半导体芯片
middle class	
	在线搜索
Internet firms	
	生物计算机

Ⅲ. Read the text again and fill in the blanks with the information you have got in the text.

1. They were of __________ computers, huge, heavy, expensive and slow, as well as using much more power than today's, but they still __________ computer science, such as the concepts of stored programs, random access.
2. Integrated circuits meant that __________ and __________ were only made on a small semiconductor chip.
3. The __________ information provided by computers has changed society's privacy landscape.
4. With the development of __________, biological computers and quantum computers will __________ in near future.

Ⅳ. Choose the best one from the items given below to complete the following passage.

Early computers __1__ to solve mathematical and engineering problems. The first general purpose of __2__ electronic computer was the Electronic Numerical Integrator and Computer (ENIAC), built by J. Presper Eckert and John V. Mauchly at the University of Pennsylvania in 1946. It filled a thirty by fifty-foot room and __3__ thirty tons. The computer had 18,000 vacuum tubes which __4__ perform calculations at a rate of 5,000 additions per second. This is __5__ than any human could perform, but a great deal slower than the computers of today.

1. A. are build B. were built C. build D. built
2. A. programming B. programmatic C. programmable D. program
3. A. weighed B. weights C. weigh D. weight
4. A. are used to B. used to C. use to D. were used to
5. A. more fast B. faster C. much faster D. very faster

Text B

Computer System

A system means a group of related parts working together. A digital computer system is mainly composed of four parts: input devices, output devices, memory and central processing (CPU) which can accept, store and process data or symbols and yield output results fast under the indication of a series of instructions. After the CPU processing the information accepted by the input devices, the output devices give out the results users need.

Users hardly touch CPUs, but all of them have used the input devices. In PC systems, users often touch keyboards, mouse, input pens, touch screens, microphones and others for direct input. Regardless of their differences, they are components that make interpretation and communication between users and computer systems. The storage devices, floppy disk drives and hard disk drives are commonly used for indirect input. Users have also employed a variety of output devices such as monitors, printers and plotters which get the outcomes from the CPU—changing the code into the forms which the users can understand. Hard disk drives or floppy disk drives often record the results on disks for next or another machine input. Some people prefer to call input/output devices, monitors, printers, plotters as peripheral devices.

Traditionally, people accept that there are two kinds of computers: digital computers and analog computers. The digital computers deal with numbers and symbols; the analog computers are only concerned with quantities, such as electrical currents or voltages. But nowadays, the former will be more powerful and occupy the positions of the latter.

 **Vocabulary**

compose [kəm'pəuz]	*v.*	组成，构成
device [dɪ'vaɪs]	*n.*	装置，装备
process ['prəuses]	*v.*	加工；处理
symbol ['sɪmbl]	*n.*	标志；符号
yield [jild]	*v.*	生产；获利
indication [ˌɪndɪ'keʃən]	*n.*	指示；表明
regardless [rɪ'gɑːdlɪs]	*adv.*	不管怎样，无论如何
component [kəm'pəunənt]	*n.*	成分；零件
floppy [flɒpɪ]	*n.*	磁碟，磁盘
interpretation [ɪntɜːprɪ'teɪʃ(e)n]	*n.*	解释，说明
indirect [ˌɪndə'rekt]	*adj.*	间接的

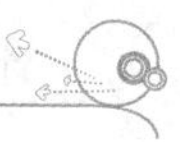

plotter ['plɒtə(r)]	*n*.	绘图机
occupy ['ɒkjʊpaɪ]	*v*.	占领
code [kəʊd]	*n*.	密码
peripheral [pə'rɪfərəl]	*adj*.	外围的；次要的
analog ['ænəlɒg]	*adj*.	模拟的
current ['kɜːrənt]	*n*.	电流
voltage ['vəʊltɪdʒ]	*n*.	电压，伏特数

Exercises

Ⅰ. Read the passage above and decide whether the following statements are true (T) or false (F).

( ) 1. A digital computer system is mainly composed of five parts.

( ) 2. Central processing (CPU) can accept, store and process data or symbols and yield output results fast under the indication of a series of instructions.

( ) 3. The storage devices, floppy disk drives and hard disk drives can't be used for indirect input.

( ) 4. Hard disk drives or floppy disk drives often record the results on the CPU for next or another machine input.

( ) 5. The analog computers deal with numbers and symbols.

Ⅱ. Match the words or phrases on the left with their meanings on the right.

1. process
2. yield
3. symbol
4. indirect
5. component
6. interpretation
7. code
8. regardless
9. device
10. occupy

A. object that refers to sth. else; emblem
B. paying no attention to sb. or sth.
C. take up or fill
D. deal with a document officially
E. thing made or adapted for a special purpose
F. not going in a straight line
G. pre-arrange signals used to send message by machine
H. bear, produce, or provide a result
I. explanation
J. any of the parts of which sth. is made

Ⅲ. The following is a list of terms of computer. After reading it, you are required to find the items equivalent to those given in Chinese in the table below.

( ) A—calculator	1. 个人计算机
( ) B—digital computer	2. 数据处理
( ) C—feedback	3. 指令

( ) D—terminal　　4. 地址
( ) E—keyboard　　5. 终端
( ) F—personal computer　　6. 键盘
( ) G—data processing　　7. 数字计算机
( ) H—instruction　　8. 反馈
( ) I—item　　9. 计算器
( ) J—address　　10. 项目，项

Ⅳ. Translate the following paragraph into Chinese.

Input Devices: Input devices are equipment that translates data and programs that humans understand into a form that the computer can process. The most common input devices for microcomputers are the keyboard and the mouse. The keyboard on a computer looks like a typewriter keyboard, but it has additional specialized keys. A mouse is a device that typically rolls on the desktop. It directs the insertion point or cursor on the display screen. A mouse has one or more buttons for selecting commands. It is also used to draw figures.

Further Reading

Cyber-Stepmother

"Stepparent" is a term we attach to men and women who marry into families where children already exist. It is most certainly a giant "step", but one does often doubt whether the term "parent" truly applies—at least that's how I used to feel about being a stepmother to my husband's four children.

My husband and I were together for six years, and with him I watched his young children become young teenagers. Although they lived primarily with their mother, they spent a lot of time with us. Over the years, we all learned to become more comfortable with each other, and to adjust to our new family arrangement. We enjoyed holidays together, ate family meals, worked on homework, played baseball and rented videos. However, I continued to feel like I was on the outside looking in, coming uninvited to a foreign kingdom.

When the children moved to a town five hours away, my husband was understandably upset. In order to continue regular communication with the kids, we set up an e-mail and chat-line service right away. This technology, combined with the telephone, enabled us to correspond with them on a daily basis by sending frequent notes and messages, and even chatting together when we were all online.

One sad thing, though, was that these modern tools of communication could still

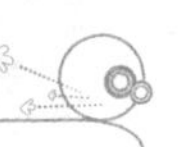

make me feel out of touch. If a computer message came addressed to "Dad", for example, I'd feel forgotten and neglected. If my name appeared along with his, it would brighten my day and make me feel like I was part of their core family unit. Yet, there was always a gap—some distance to be crossed—not just over the telephone wires.

Late one evening, as my husband was sitting in front of the television and I was catching up on my e-mail, an "instant message" appeared on the screen. It was Margo, my oldest stepdaughter, also up late and sitting in front of her computer five hours away. As we had done in the past, we sent several messages back and forth, exchanging any current news.

When we would "chat" like that, she wouldn't necessarily know if it was her dad or I on the other end of the keyboard unless she asked. That night, she didn't ask and I didn't identify myself either. After hearing about the latest fashions at the mall, details about a dance at her school, and a history project that was in the works, I commented that it was late and I should get to sleep. Her return message read, "Okay, talk to you later! Love you!"

As I read this message, a wave of sadness ran through me as I realized that she must have thought she was writing to her father the whole time. She and I would never have openly exchanged such words of affection. Feeling guilty for not identifying myself, yet not wanting to embarrass her, I simply responded, "Love you too! Have a good sleep!"

I thought again of their family circle and felt the sharp ache of emptiness I thought I had grown accustomed to. Then, just as my fingers reached for the keys to return the screen to black, Margo's final message appeared. It read, "Tell Dad good night for me too." With tear-filled eyes, I turned the machine off.

Writing

Understanding and Filling in Registration Forms

当人们参加各种会议、活动或是入住酒店时，经常需要填写登记表。一般情况下，登记表包含填写人的基本信息如下：称呼、姓、名、出生日期、家庭住址、联系方式、国籍等。

例如：

1. Title：称呼，即称先生、小姐、女士、夫人还是某某博士。
2. Family name (Surname or Last name)：姓。
3. Given name (First name or Christian name)：名。
4. Date of birth：出生日期。英国人习惯按日/月/年（dd/mm/yyyy）的顺序而美国人习惯按月/日/年（mm/dd/yyyy）的顺序。

5. Home address：家庭地址。注意顺序，从门牌号、街道名、城市名到国家名、邮政编码，从小到大依次填写。
6. Country of birth：出生国，即在哪个国家出生。
7. Nationality：国籍，即现在是哪个国家的公民。注意：此处应用国名的形容词形式，如例文中用到 American 而不是 America 或 U. S. A. 。中国国籍用 Chinese 而不用 China。

Sample

Registration		
Title：Miss（Miss/Ms. /Mr. /Mrs. /Dr.） **Given name（First name）**：Joan		**Family name（Surname）**：Brown
Date of birth： 03 Day 06 Month 1983 Year		
Home address：No. 107 Selden Street，San Diego，California USA 92117		
Tel：（414）895-2387	**Fax**：（414）895-2386	**E-mail**：jbrown@sina. com
Country of birth：Australia	**Nationality**：American	

Task：The following is a registration form. Read it and then complete the sentences.

Family Name	Jameson	Given Name	Henry
Street	5th Ave.	City	New York
State	New York	Zip Code	10276-0906
Tel（H）	（44）1902-32123	Tel（O）	（44）1902-32323
E-mail	hjameson@google. com	Student ID	F20089595

1. Who filled out the form?
   Mr. ________ filled it out.
2. Henry lives at ________________.
3. When you are in New York and Henry is at school，please dial ________ if you want to talk to him.
4. If you are in China and Henry is at home，please dial ________ when you want to talk to him.
5. If you want to reach Henry by computer，use ________________.

Unit 2 Product Design & Operating

Learning Objectives

In this unit you should get familiar with:

- Apple's chief designer: Jonathan Ive;
- Apple's product design;
- The operating system;
- The format of memos.

Speaking

Dialogue 1

A: Could you give me the size of the printer?

B: Yes. It's 15 centimeters high.

A: How about its width?

B: It's 30 centimeters wide.

A: How long is it?

B: The length is 65 centimeters.

A: By the way, how much does it weigh?

B: Its weight is 5 kilograms.

Dialogue 2

A: What's the important part of the computer?

B: The most important part of your computer isn't the hard disk or the monitor or the printer. The most important part is the data you use.

A: Why?

B: It's the only part of your computer that can't be replaced. If you don't make a back-up copy of it and something bad happens to your computer, you will never see your data again.

A: I should buy a back-up tape drive.

B: That's a very good idea.

Dialogue 3

A：Is this the Telecommunication Bureau?

B：Yes，this is the New-Service Demonstration Center. What can I do for you?

A：I want to apply for an Internet account. Can you make house calls?

B：Certainly. Leave us your address，please?

A：OK. Will I have to wait for a long time?

B：We will arrange this in 24 hours.

A：By the way，is the service charge high?

B：Oh，no. It's free of charge.

A：What else do you want?

B：Please tell us your telephone number so that we can contact you.

A：It is 8765432.

B：Now all you need to do is just wait at home. See you later.

A：Bye-bye.

Dialogue 4

A：Hello，is that the Data Center?

B：Yes. What's the matter?

A：I am one of your clients who want to thank you for your good service.

B：You are welcome. Why are you so glad?

A：I have changed my Internet Bar from dial-up into ASDL after your convincing recommendation.

B：Isn't the result satisfying?

A：Yes. It was very slow when 20 people went online；but now it is all right when 50 people get on line.

B：Thanks a lot for your support to our work and for your feeding back so good information.

Dialogue 5

A：I need some information，but I don't know which web site to search.

B：You can use the search function of the browser to do it.

A：How to operate it?

B：You can click the search button in the browser and then put the keywords in the input box.

A：Any way else?

B：You can also use the special search web sites，such as Yahoo and Sohu，

commonly named as the search engines.

A: Only the search engines web sites have the search function, isn't it?

B: No. Many famous web sites also have the search function, and that goes for Netease.

A: They have the same operation method, don't they?

B: That's right. You can try after back home.

A: That's all. Thanks.

Dialogue 6

A: Good morning. May I ask you some questions about my computer?

B: Good morning. What can I do for you?

A: The system in my computer has crashed. I have tried several times to reboot it, but it doesn't work. What can I do?

A: How did that happen?

B: The screen went blank while I was just drawing with Photoshop. It all happened suddenly, and I am totally at a loss why.

Text Reading

Text A

Jonathan Ive[1]: The Innovative Architect behind Apple's Greatest Creations

Concealed from both the public and internal employees, Apple projects perfection by hiding its process from the world.[2] Much like its next-generation electronics, the company's master of industrial design, Jonathan Ive, is also tucked away from the public eye.

With over three hundred design patents credited to his name, Ive's innovative techniques and methods have paved the way for the "Post-PC"[3] world. His scrupulous attention to detail has moved the consumer electronics industry forward. Alongside his close friend, Steve Jobs, he has played an extremely important role in reviving Apple, yet his participation in creating the innovative products Apple is so well-known for has largely gone unnoticed.

For three years, Ive was forced to design Newton devices and printer trays, most of which never even reached market. He even forged the first flat panel computer, yet its price tag immediately condemned it in the eyes of consumers. With no apparent way to utilize his talents, Jonathan looked for a way out of the United States.

His pessimism was quickly changed to hopeful optimism with the return of Steve

Jobs, who came across the designer's prototypeson a tour around the company's campus. The now iconic leader of one of the largest companies in the consumer electronics industry demanded that Ive be promoted, tasking him with forging Apple's future with the iMac.[4]

Characterized by its curvy, translucent backing that differentiated it from every computer in its class, the iMac was instantly a hit. Although it might not have met Jobs' expectations completely, the mouse and choice of USB technology causing his pain, it fit the bill, providing consumers with a cheap cathode-ray-tube[5] computer. It was a step forward that saved Apple from being swallowed by financial stress.

Ive's next project was the iPod, a 2001 media player that merged a simple form factor with intuitive controls and data transfer translating to massive success. The iPod line was shocked life back into Apple and made it an industry leader in multimedia ventures. This eventually resulted in Ive's work on the iPhone and later the iPad.

The acclaimed industry innovator's unique approach targets the user as opposed to the device. In introducing the iPad, the acclaimed designer and innovator shared his most important principle of design, which is to have the user define the device rather than have the device define the user. This sentiment has changed the technology universe forever.

Ive played an important role in creating the current tablet market, which did not exist in its current form before the introduction of the mind-bogglingly simple iPad. He changed the digital music industry by forging the iPod's candy bar form factor. He stretched the abilities of aluminum unibodies to the limit with recent MacBook models and the iPad, making it an Apple's signature material.

Recognized for his part in creating Apple's products in 2010, *Fortune* magazine named Ive "World's Smartest Designer". He has been showered with industry-specific awards for his work on the iPhone, iPad, MacBook, iMac, and iPod. One of his crowning achievements did not come in the form of an award, but rather when the British monarch revealed her adoration of the iPod in 2005.

While Ive's work process might not be entirely known, rest assured that he is in his workshop with his partners, paving the path of the computer industry. Though we will never understand how Ive works, his work is undoubtedly "magical".

Notes:

[1] Jonathan Ive：乔纳森·艾维。现任 Apple 公司设计师兼资深副总裁。他曾参与设计了 iPod、iMac、iPhone、iPad 等众多苹果产品。

[2] Concealed from both the public and internal employees, Apple projects perfection by hiding its process from the world.

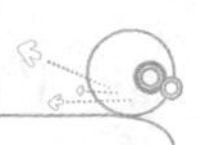

苹果公司的产品研发项目常常被隐藏在公众和内部员工的视线之外，其通过对研发过程的严格保密展现了产品的完美无缺。

在这里，project 一词为动词，意为“生动地表现，展现，设计”。

[3] Post-PC：后个人计算机时代。以公元 2000 年作为科技史的一个分水岭，那么公元 2000 年之前可以称之为“PC”（Personal Computer）时代；而公元 2000 年之后则被称为“后 PC”（Post-Personal Computer）时代。后 PC 时代是指将计算机、通信和消费产品的技术结合起来，以 3C 产品的形式（因为以上三者英文都是以“C”字母开头的）通过 Internet 进入家庭。

[4] iMac：是一款苹果计算机，针对消费者和教育市场的一体化苹果 Macintosh 计算机系列。iMac 的特点是它的设计。早在 1998 年苹果总裁斯蒂夫·乔布斯就将“what's not a computer”（不是计算机的计算机）这一概念应用于设计 iMac 的过程。结果造就了软糖——iMac G3，台灯——iMac G4 和相框——iMac G5。由于 iMac 在设计上的独特之处和出众的易用性，它几乎连年获奖。

[5] cathode-ray-tube：阴极射线管，由德国物理学家 Kari Ferdinand Braun 发明。该射线管于 1897 年被用于一台示波器中，首次与世人见面。阴极射线管是将电信号转变为光学图像的一类电子束管，主要由电子枪、偏转系统、管壳和荧光屏构成。

Vocabulary

conceal [kən'siːl]	*v.*	隐藏；隐瞒
project [prə'dʒekt]	*v.*	生动地表现，展现
tuck [tʌk]	*v.*	使隐藏
patent ['pætnt]	*n.*	专利
innovative ['ɪnəvetɪv]	*adj.*	革新的，创新的
scrupulous ['skrʊpjələs]	*adj.*	细心地，一丝不苟的
revive [rɪ'vaɪv]	*v.*	振兴，恢复
condemn [kən'dem]	*v.*	指责，责备
prototype ['prəʊtə'taɪp]	*n.*	模型
iconic [aɪ'kɒnɪk]	*adj.*	偶像的
forge [fɒrdʒ]	*v.*	锻造，打制
curvy ['kɜːvɪ]	*adj.*	曲线美的
translucent [træns'lʊsnt]	*adj.*	半透明的
differentiate [dɪfə'renʃeɪt]	*v.*	区分，区别
merge [mɜːdʒ]	*v.*	合并，融合
intuitive [ɪn'tuɪtɪv]	*adj.*	易于理解和使用的
acclaimed [ə'kleɪmd]	*adj*	受到赞扬的
unique [jʊ'nɪk]	*adj.*	独特的
define [dɪ'faɪn]	*v.*	给……下定义
sentiment ['sentɪmənt]	*n.*	观点

mind-bogglingly ['maɪnd,bɒglɪŋli]	*adv*.	令人难以置信地
unibody ['juːnɪ'bɒdi]	*n*.	一体式结构
shower ['ʃaʊə]	*v*.	大量地给予
monarch ['mɒnək]	*n*.	帝王
adoration [ˌædə'reɪʃən]	*n*.	崇拜，爱慕

Exercises

Ⅰ. Answer the following questions briefly according to the text.

1. When did Ive change his pessimism to hopeful optimism?
2. What is the special feature of iMac designed by Ive?
3. How did Ive help Apple become an industry leader in multimedia ventures?
4. What is Ive's design principle?
5. How should we describe Ive's work?

Ⅱ. Fill in the table below by giving the corresponding Chinese or English equivalents.

design patent	
	后个人计算机时代
Newton device	
	打印机纸盘
flat panel computer	
	设计模型
cathode-ray-tube computer	
	多媒体播放器
unique approach	
	设计原则

Ⅲ. Read the text again and fill in the blanks with the information you have got in the text.

1. ________ both the public and internal employees, Apple projects perfection by hiding its process from the world. Much like its ________, the company's master of industrial design, Jonathan Ive, is also ________ the public eye.
2. With no apparent way to ________, Jonathan looked for a way out of the United States.
3. Although it might not have ________ completely, the mouse and choice of USB technology ________, it fit the bill, providing consumers with a cheap cathode-ray-tube computer.
4. Ive ________ creating the current tablet market, which did not exist in its current form before the introduction of the ________ iPad.

5. One of his _________ did not come in the form of an award, but rather when the British monarch _________ of the iPod in 2005.

Ⅳ. **Translate the following paragraph into Chinese.**

The entrepreneur owns a big company with more than two thousand employees. He likes using Google to find information for his international trade. His secretary often helps him with correspondence, meetings, and so on. He goes to all kinds of breakfast meetings, which rarely last more than one hour. In his eyes, polite dining at the table is very important. The industrial designers in his company are often asked by him to do their best to make products attractive.

Text B

Operating System

An operating system is a program, which acts as an interface between a user of a computer and the computer hardware. The purpose of an operating system is to provide an environment in which a user may execute programs. The primary goal of an operating system is thus to make the computer system convenient to use. A secondary goal is to use the computer hardware in an efficient manner.

An operating system is similar to a government. Its hardware, software, and data provide the basic resource of a computer system. The operating system provides the means for the proper use of these resources in the operation of the computer system. Like government, the operating system performs no useful function by itself. It simply provides an environment within which other programs can do useful work.

We can view an operating system as a resource allotter. A computer system has many resources (hardware and software) which may be required to solve a problem: CPU time, memory space, file storage space, input/output (I/O) devices, and so on. The operating system acts as the manager of these resources and allocates them to specific programs and users as necessary for their tasks. Since there may be many, possibly conflicting, requests for resources, the operating system must decide which requests are allocated resources to operate the computer system fairly and efficiently.

Operating systems and computer architecture have had a great deal of influence on each other. To facilitate the use of the hardware, operating systems were developed. As operating systems were designed and used, it became obvious that changes in the design of hardware could simplify the operating system. In this short historical review, notice how the introduction of new hardware features is the natural solution to many operating system problems.

Vocabulary

interface [ˈɪntəfeɪs]	*n*.	界面，<计>接口
execute [ˈeksɪkjuːt]	*v*.	实行，执行
primary [ˈpraɪmərɪ]	*adj*.	主要的；基本的
secondary [ˈsek(ə)nd(ə)rɪ]	*adj*.	第二的；次要的
resource [ˈrɪsɒrs]	*n*.	资源
view [vju]	*v*.	观察；查看
allotter [əˈlɒtə]	*n*.	分配器
allocate [ˈæləkeɪt]	*v*.	分配
conflict [ˈkɒnflɪkt]	*v*.	冲突；争执
architecture [ˈɑːkɪtektʃə]	*n*.	建筑学
facilitate [fəˈsɪlɪteɪt]	*v*.	促进；使容易
simplify [ˈsɪmplɪfaɪ]	*v*.	简化；使简易
historical [hɪˈstɒrɪk(ə)l]	*adj*.	基于历史的

Exercises

Ⅰ. Read the passage above and decide whether the following statements are true (T) or false (F).

(　) 1. An operating system is a program, which acts as an interface between a user of a computer and the computer software.

(　) 2. The secondary goal of an operating system is to make the computer system convenient to use.

(　) 3. An operating system's hardware, software, and data provide the basic resource of a computer system.

(　) 4. Operating system cannot decide which requests are allocated resources to operate the computer system fairly and efficiently.

(　) 5. The introduction of new hardware features can solve many operating system problems.

Ⅱ. Complete the following sentences by translating the Chinese in the brackets.

1. Office automation does not always mean ________ (按一下按钮就可以把一切从头做到尾).
2. Programming too much into the work process in the first round of automation can sometimes ________ (使整个系统不够灵活) and stop new ideas from coming to the surface.
3. Office automation may also be limited by ________ (用户使用程序的水平).

4. Some prefer Excel instead of Access because of their familiarity and because they are afraid that if ________ (损坏了 Access 数据库), they will not be able to fix it themselves.
5. ________ (对可利用的工具做出明智的选择) may be all that is necessary.

Ⅲ. Choose the best one from the items given below to complete the following passage.

Another interesting __1__ on the Internet is the electronic bulletin board which is also called Bulletin Board System, BBS for short. It allows __2__ to post and retrieve __3__ that are not directed to a specific user, much like announcements are posted on an office __4__. BBS has been used for everything from dating service and want ads to highly specialized applications such as the exchange of research __5__ in a narrow scientific field.

	A	B	C	D
1.	A. type	B. application	C. difference	D. advantage
2.	A. users	B. students	C. businessmen	D. programmers
3.	A. experiences	B. presents	C. books	D. messages
4.	A. window	B. desk	C. bulletin board	D. door
5.	A. data	B. lists	C. teachers	D. jobs

Ⅳ. Translate the following paragraph into Chinese.

The most important program that runs on a computer is the operating system. Every general-purpose computer must have an operating system to run other programs. The purpose of an operating system is to perform basic tasks, such as recognizing input from the keyboard, sending output to the display screen, keeping track of files and directories on the disk, and controlling peripheral devices such as disk drives and printers.

Further Reading

The Magic 3D-Printing Pen

Humans are accustomed to drawing in the air. We gesture with our hands when talking and will try to illustrate charade secrets by "drawing" objects in space. 3D-printing pens takes those gestures, makes them tangible and, in the hands of an artists, beautiful. Recent 3D-printing pens have been cool, but clunky affairs. LIX Pen, however, is something different. It's light, small and apparently needs no more power than you can draw from your run-of-the-mill laptop. Now it's coming to Kickstarter.

Measuring 6.45 inches long, 0.55 inch in diameter and weighing just 1.23 ounces, the aluminum 3D-printing pen (which also comes in black) really is pen sized. You hold it just like a pen, and plug a 3. 5mm-like jack into the base and the other end of your cable into your computer. The juice allows LIX to heat to over 300-degrees Fahrenheit, though the plant-based PLA filament (it can also use the stronger ABS plastic) only needs to heat to 180-degrees to work. That filament is fed in through a hole in the base and emerges as a super-heated liquid on the tip so you can start doodling in the air.

Unlike 3D printers, there is no program guiding the printing tip. Instead, to create 3D objects, you simply start drawing in the air with the LIX Pen, moving slowly as the melted filament draws out. It cools quickly so that your structure remains rigid. Each filament rod is about 10 centimeters long and should, according to the company, last for about two minutes of air-drawing.

3D-printing pen works for everything from abstract sculpture to fine art and jewelry to T-shirt design. The only limit, it appears, is your skill level and ability to hold and move the pen very, very steadily.

LIX co-founder Anton Suvorov, told Mashable the company's 3D-printing pen "has no concurrence on the market", and it should arrive in Kickstarter sometime around April 14, where the company will be taking pre-orders. The starting price, at least for the campaign, will be $139. 95. LIX also sells, for $59. 95, a ballpoint pen replica of LIX that is nothing more than a regular pen, but why would anyone want that?

Writing

Memos

备忘录（Memos）是公司内部最基本、最常用的一种信息传递方式，如会议安排、情况报告、责任确认、问题处理等。备忘录通常由公司统一印制成标准的表格。

在备忘录中，To 一栏的收笺人和 From 一栏的发笺人可以是姓名后加上其职位和部门，也可只写职位。收笺人的姓名前可写上 Mr. ，Mrs. ，Ms. ，Dr. 等，而发笺人的姓名前则不写。

Subject 即内容标题，一般用名词或动名词词组等少数几个词做简略叙述。

内容部分是备忘录的主要部分，应力求简明，确切。

最后，结束时不用签名。

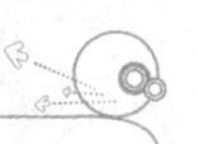

Sample

<table>
<tr><td>

MEMORANDUM

To: Mr. Lin, Regional Sales Manager
From: Assistant Sales Manager
Date: October 14th, 2008
Subject: Handling of the Enquiry

I have dealt with the enquiry that you passed to me on Friday. The enquiry was from Mr. E. King, who wanted to know whether we could offer him a special wholesale discount. I told Mr. King that we could offer his company a 5% discount on orders over $1,000.

</td></tr>
</table>

Task: Write a memo according to the information given in Chinese. Some parts have been done for you.

2008年7月5日，罗锐向质量控制部经理（Quality Control Manager）汇报：在检测新机器时发现该机器有问题。他建议新机器停止生产以保证产品质量。

<table>
<tr><td>

MEMORANDUM

To: 1) ________________
From: Luo Rui
Date: 2) ________________
Subject: 3) ________________

At the recent test, I discovered that 4) ________________ (新机器有问题). May I suggest that the production of the new machine be stopped to ensure 5) ________________ (产品质量)?

</td></tr>
</table>

Unit 3 Establishing Business Relation

Learning Objectives

In this unit you should get familiar with:

- Establishing business relation about computers;
- Apple company;
- iPad Air's feature;
- Apple's computer lineup;
- Business card.

Speaking

Dialogue 1

A: Hi, Sam. How's it going?

B: Pretty good. I'm going to buy a new computer this afternoon.

A: What kind are you going to buy?

B: I think I'll buy a desktop, maybe a Dell.

A: They are very popular in America, and the prices are really coming down.

Dialogue 2

A: Can I help you?

B: Yes, I need to buy a computer for this semester. I was told it is cheaper to buy computers here.

A: Well, you heard right. You can get an excellent deal on a new computer here. We have great discounts for students.

B: I don't know much about computers. But I know I want a desktop computer with a lot of memory. And I need a printer.

A: Well, first let's consider your computer. Here, for example, is a system I highly recommend—the Power Macintosh G3. It comes with 512M of total memory.

Dialogue 3

A: May I help you?

B: Yes, I want to buy a new computer.

A: How much RAM do you need? How big a hard drive will you need?

B: Well, Windows XP needs at least 256 MBS RAM, and I'll be using a lot of word processors and game programs.

A: I recommend a Pentium IV with an 80 GB hard drive.

Dialogue 4

A: Hi, Han Mei!

B: Look! My parents bought a computer for me.

A: Great! Can I see it?

B: Sure! But I don't know how to use it yet. I have to learn to use it all by myself. My parents don't know much about computers.

A: Can you teach yourself?

B: My cousin is working at a computer company. She's a computer engineer. She can help me.

Dialogue 5

A: Hey, Sam, where have you been this week?

B: I was writing a web page for my new business—selling Bizet pottery.

A: Cool. What did you put on the page?

B: I put a CGI form on it so people could send me information.

A: How many hits has it gotten?

B: It's getting about 100 a day.

A: Have you sold anything?

B: Not yet…

Dialogue 6

A: Good morning. Lenovo Company. This is Susan Packer speaking. What can I do for you?

B: Good morning. This is Paul Lee, purchasing manager from Tarmacs Company. I'd like to talk with your sales manager to learn about your enterprise laptops targeted at small-to-medium businesses.

A: I'm sorry. Our sales manager is occupied at the moment. Would you like to make an appointment?

B: Yes, that'll be great.

A: How about 10 o'clock tomorrow morning?

B: OK. Let's make it at 10.

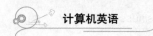

Text Reading

Text A

Apple's Favorite

—iPad Air

Apple Inc. is refreshing its iPad lineup and slashing the price of its Mac computers ahead of the holiday shopping season, as it faces an eroding tablet market share and growing competition from rival gadget makers. [1]

Apple unveiled a new, thinner, lighter tablet called the "iPad Air"[2] along with a slew of new Macs Tuesday at an event in San Francisco. The iPad Air weighs just 1 pound, compared with 1.4 pounds for the previous version. Apple marketing chief Phil Schiller called the tablet a "screaming fast iPad". He said it is eight times faster than the original iPad that came out in 2010.

The iPad Air will go on sale Nov. 1 and start at $499 for a model with 16 gigabytes of memory. [3] Apple plans to phase out its third and fourth generation iPads while the iPad 2, which launched in 2011, continues selling at $399. A new iPad Mini, meanwhile, will be available later in November starting at $399 for a 16-gigabyte model. It has a retina display designed to give it a clearer, sharper picture and the same 64-bit chip that powers the iPad Air. [4]

Apple also refreshed its computer lineup. A new, 13-inch Mac Book Pro[5] with Retina display is thinner and lighter, Schiller said, adding that the laptop has up to 9 hours of battery life, enough to watch the entire trilogy of *The Dark Knight*[6] on one charge. The notebook's new price is lower: $1, 299, compared with $1, 499 for the previous version.

A larger Mac Book Pro, with a 15-inch monitor and 256 gigabytes of storage starts at $1, 999, compared with $2, 199 for the previous version.

The Mac Pro, a high-end desktop computer aimed at what Apple calls "power users", will be available in December for $2, 999.

The company also said that its latest computer operating system, Mavericks, is available free of charge.

Notes:

[1] Apple Inc. is refreshing its iPad lineup and slashing the price of its Mac computers ahead of the holiday shopping season, as it faces an eroding tablet market share and growing competition from rival gadget makers.

苹果公司赶在圣诞节购物旺季之前推出了 iPad 系列的新品，削减了 Mac 计算机的价

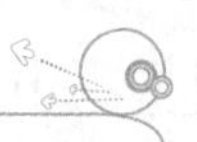

格。这些措施都是苹果公司面对其在平板电脑市场上缩小的市场份额，和来自其他平板制造商的日益剧烈的竞争压力所做出的努力。

[2] iPad Air：美国苹果公司于北京时间2013年10月23日举行发布会，正式公布了全新的iPad Air。iPad Air最大的变化是其整体设计，更偏向于mini，使得iPad Air拥有令人难以置信的纤薄轻巧。得益于铝金属Unibody一体成型设计，机身仅有7.5 mm之薄。体积比上一代iPad减少近四分之一，重量不足500 g，坚固程度也同样令人难以置信。它还配备了绚丽夺目的Retina显示屏，凭借2048×1536像素分辨率和超过310万的像素数量，照片和视频可精细到纤毫毕现，文字也显得清晰明锐。

[3] The iPad Air will go on sale Nov. 1 and start at $499 for a model with 16 gigabytes of memory.

iPad Air将于11月1日起发售，初始售价为499美元，内备有16G的内存卡。

[4] It has a retina display designed to give it a clearer, sharper picture and the same 64-bit chip that powers the iPad Air.

这款迷你iPad配备有画质清晰细致的视网膜屏幕，以及和iPad Air上配备的相同的64字节的芯片。

[5] Mac Book Pro是款苹果公司用来取代PowerBook G4产品线的英特尔核心的笔记型计算机。这款迷你iPad配有画质清晰细致的视网膜屏幕，以及和iPad Air上配备的相同的64位的芯片。

[6] the trilogy of *The Dark Knight*：《蝙蝠侠：黑暗骑士》三部曲系列电影。包括三部作品：《蝙蝠侠：开战时刻》《蝙蝠侠：黑暗骑士》和《蝙蝠侠：黑暗骑士崛起》。

Vocabulary

refresh [rɪ'freʃ]	*v*.	更新；使……恢复
lineup ['laɪnʌp]	*n*.	阵容；一组人
slash [slæʃ]	*v*.	削减；斜线；猛砍
eroding [ɪ'rəudɪŋ]	*adj*.	侵蚀的；侵害的
tablet ['tæblət]	*n*.	写字板；小块
share [ʃeə]	*n*.	份额；股份
gadget ['gædʒɪt]	*n*.	小玩意；小器具
original [ə'rɪdʒənl]	*adj*.	以前的；原始的
gigabyte ['gɪgəbaɪt; 'dʒ-]	*n*.	十亿字节，十亿位组
launch [lɒntʃ]	*v*.	发射；发行
trilogy ['trɪlədʒɪ]	*n*.	三部曲；三部剧
charge [tʃaːdʒ]	*n*.	电荷；电池；费用
previous ['priːvɪəs]	*adj*.	原始的；最初的；独创的
version ['vəːʃ(ə)n]	*n*.	版本
unveil [ʌn'veɪl]	*v*.	公开；原形毕露
screaming ['skriːmɪŋ]	*n*.	尖叫的

Phrases and Expressions

a slew of　　大量的
phase out　　逐步淘汰
retina display　　视网膜显示屏

Exercises

Ⅰ. **Each IT company has several departments. Match each department with its main responsibility.**

1. Research & Development　　2. Production　　3. Testing
4. Marketing & Sales　　5. After-Sales Service
6. Human Resources　　7. Finance　　8. Purchasing

(　) A. It deals with bills, salaries, taxes, investment and budgets, etc.
(　) B. It is responsible for advertising and market research and it organizes the selling of the products.
(　) C. It produces the products.
(　) D. It deals with staff and is responsible for recruitment and staff training.
(　) E. It is responsible for buying the materials the company needs to make products.
(　) F. It conducts researches, develops new products and improves the finished products.
(　) G. It puts the finished products into a series of tests to see their overall performance.
(　) H. It solves the problems that come from the customers.

Ⅱ. **Answer the following questions briefly according to the text.**

1. Which actions did Apple take to face competition?
2. What advantages did iPad Air have?
3. What are the functions of a new iPad Mini?
4. How long does the new laptop's battery life up to?
5. Is Mac Pro's latest computer operating system free to the customers ?

Ⅲ. **Read the text again and fill in the blanks with the information you have got in the text.**

1. Apple unveiled a new, thinner, lighter tablet called the "iPad Air" ________ new Macs Tuesday at an event in San Francisco.
2. He said ________ the original iPad that came out in 2010.
3. The iPad Air weighs just 1 pound, compared with 1.4 pounds for ________.
4. The company also said that its latest computer operating system, Mavericks, ________ free of charge.

Ⅳ. **Translate the following paragraph into Chinese.**

Nobody else in the computer industry, or any other industry for that matter, could put on a show like Steve Jobs. His product launches, at which he would stand alone on a black stage and conjure up a "magical" or "incredible" new electronic

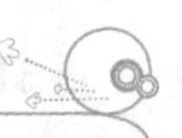

gadget in front of an awed crowd, were the performances of a master showman. All computers do is fetch and shuffle numbers, he once explained, but do it fast enough and "the results appear to be magic". He spent his life packaging that magic into elegantly designed, easy to use products.

Text B

A Further Inquiry about Laptops

Dear Lucy,

After talking with your sales manager Mr. Lee both in person and on the phone about our expectations of your laptops, we are basically satisfied with your recommendation of Mac Book Pro.

As you know, we are a company featuring in career training. Your cutting-edge technologies such as thinner and lighter 13-inch Mac Book Pro with retina display, and a larger Mac Book Pro with a 15-inch monitor and 256 gigabytes of storage, and your latest computer operating system, excellent multimedia capabilities, as well as stylish design are just fit for our needs. Besides, the software and service packages that come with your computers seem to be developed with users like us in mind. With those packages, we can get timely help when we have trouble in things such as connecting to the Internet, managing passwords, data recovery and updating the computers.

However, after examining your price list and carefully comparing it with that offered by other laptop vendors, we feel that your price for the Mac Book Pro is a little bit high. So, I am writing to you about the possible discount you might give when we make a volume purchase. If we buy 100 at one time, could you give us 10 percent off?

I am also writing to you to inquire of your LAN equipment, for we are going to build a LAN in our company pretty soon. Could you please send us some brochures of your products so that we can have a look first?

Look forward to hearing from you soon.

Yours sincerely,

Tony Zhang

Vocabulary

inquiry [ɪn'kwaɪrɪ]	*n.*	咨询，询问
expectation [ekspek'teɪʃ(ə)n]	*n.*	期待；预期；指望
recommendation [ˌrekəmen'deɪʃ(ə)n]	*n.*	推荐；建议；推荐信
cutting-edge [ˌkʌtɪŋ'edʒ]	*adj.*	先进的，尖端的

capability [keɪpəˈbɪləti]	*n*.	能力
stylish [ˈstaɪlɪʃ]	*adj*.	时髦的；现代风格的；潇洒的
timely [ˈtaɪmli]	*adj*.	及时的；适时的
update [ˌʌpˈdeɪt]	*v*.	更新；升级
vendor [ˈvendə]	*n*.	供应商，销售商
discount [ˈdɪskaʊnt]	*n*.	折扣
brochure [ˈbrəʊʃə; brɒˈʃʊə]	*n*.	手册，小册子

Exercises

Ⅰ. Read the letter again and choose the best answer to each question.

1. According to the e-mail, what is NOT true about Mr. Lee and Tony Zhang ?
   A. They have met each other face to face.
   B. They have talked with each other on the phone.
   C. They have discussed about Apple's computer lineup.
   D. They have signed a contract for the sale of Apple's computer lineup.
2. What is Tony Zhang's attitude toward Apple's Mac Book Pro ?
   A. Extremely satisfied.　　B. A little bit satisfied.
   C. Generally satisfied.　　D. Not satisfied at all.
3. According to the letter, what is NOT true about the Apple's Mac Book Pro?
   A. Lighter and larger.　　B. With Retina display.
   C. 256 gigabytes of storage.　　D. They are rather cheap.
4. What is Tony Zhang's purpose in writing this letter?
   A. To ask about a possible discount for a large purchase and LAN equipment.
   B. To tell Mr. Lee his satisfaction with their computer lineup.
   C. To inform Mr. Lee that their company is going to build a LAN.
   D. To tell Mr. Lee that they cannot afford its computer lineup.

Ⅱ. Match the following terms with their Chinese meanings.

1. computers lineup	A. 前沿技术
2. retina display	B. 数据恢复
3. career training	C. 视网膜屏幕
4. cutting-edge technology	D. 职业培训
5. multimedia capability	E. 计算机系列
6. stylish design	F. 软件和服务包
7. software and service package	G. 密码
8. password	H. 多媒体功能
9. data recovery	I. 大宗采购
10. volume purchase	J. 时尚设计

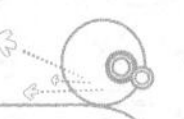

Ⅲ. Complete the following sentences by translating the parts given in Chinese.

1. ____________（随着科学技术的发展），more and more information including numerical data and a great deal of none-numerical data comes out.
2. Especially in these years，with the development of database and computer network technology，computer users in different districts and countries can ____________（通过网络通信共享信息资源）.
3. Now，as CAM，CAD and CAI being used very widely，____________（全面自动化）from design to production has fulfilled in many fields.
4. Computer is electronic equipment which can make ____________（算术和逻辑运算）and process information rapidly and automatically.
5. The technical idea underlying treacherous computing is that the computer includes ________________（一个数位加密以及签章装置），and the keys are kept secret from you.

Further Reading

Entrepreneurs

Amazon. com is the world's first and largest Internet bookstore，started by Jeff Bozos in 1995. It was serving millions of customers in 120 different countries five years later. In 1999，Jeff Bozos was selected as the *Time Magazine*'s Person of the Year，a very great honor.

Jeff Bozos is a perfect example of entrepreneurs. *Entrepreneur* means a person who starts a completely new business or industry，like Jeff Bozos，who started the very first Internet bookstore. Entrepreneurs are regarded as highly successful in American society，and in many other countries，too. You might wonder：what kind of people entrepreneurs are，and what kind of background they come from.

Let's begin by looking at the qualities of entrepreneurs. There are two qualities that all entrepreneurs have in common. First，entrepreneurs have vision. They have the ability to see opportunities that other people simply do not see. Take Jeff Bozos for example. One day in 1994，he was surfing the Internet when suddenly he had a brilliant idea：why not use the Internet to sell products? Remember，at that time，no one was using the Internet in that way.

The other quality that all entrepreneurs possess is that they are not afraid to take risks. They're not afraid to fail. For example，Frederick Smith founded Federal Express，the company that delivers packages anywhere in the United States overnight. Smith first suggested the idea for his company in a college term paper. Do you know what grade he got on it? A "C"! But this didn't stop him, and today his company is worth more than two billion dollars and employs more than 25, 000 people.

Entrepreneurs come from many different backgrounds. First of all, some entrepreneurs are well educated, like Jeff Bozos, who graduated from Princeton University. But others never even finished college. Next, some entrepreneurs come from rich families, like Fred Smith, the founder of Federal Express. In contrast, other entrepreneurs come from poor families. You may be interested to know, in addition, that many entrepreneurs are immigrants or the Children of immigrants. A great example is Andrew Grove, the former chairman of the Intel computer company, who was born in Hungary and came to America as a refugee after World War II.

The third difference is that although many entrepreneurs start their careers at a young age, lots of others don't get their start until age 40 or later. And finally, it's important to remind you that entrepreneurs are not always men. A famous woman entrepreneur, for example, is Debbi Fields, who founded the Mrs. Fields Cookie Company. You can find her shops in malls all over North America and Asia.

Entrepreneurs are among the most respected people in the United States. They are cultural heroes, like movie stars or sports heroes. Why? Because, starting with a dream and working very hard, these people created companies that solved serious, important problems. They provided jobs for millions of people, and in general their companies made life easier and more pleasant for all of us.

Writing

Writing a Business Card

在各种社交活动和商务往来中，名片是一种常用的联系方式。在名片上可以读到以下信息：公司名称、持有人的名字、称呼、职位、地址、邮政编码、电话号码、传真号码以及邮箱地址等。

Sample

Avon Company Limited (Guangzhou)

Liu Dong

Purchasing Assistant

Address: 422 Huangshi Road, Baiyun District, Guangzhou

Tel: 020-86453599　　Postal code: 510426

Fax: 020-86625598　　E-mail: ld@sina. com. cn

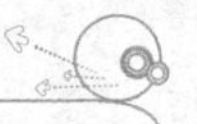

Task: Complete the following chart in English with the information given by sample. The first has been done for you.

Employment organization	Avon (Guangzhou) Company Limited
Card holder	
Title/position	
Address	
Postal code	
Fax	
Telephone	
E-mail	

Unit 4 Product Test

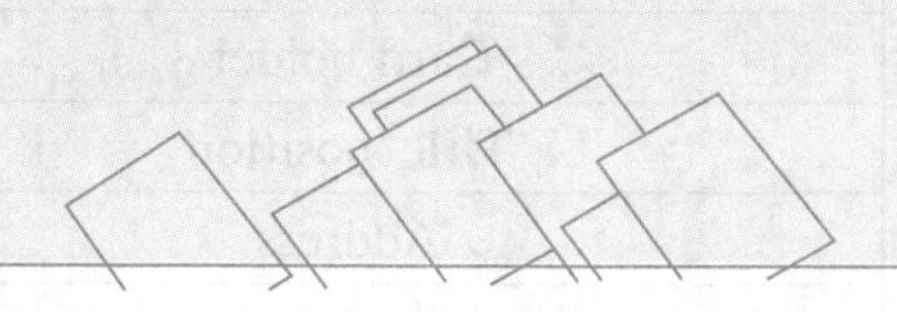

Learning Objectives

In this unit you should get familiar with:

- The process of web accessibility testing;
- The purposes of software testing;
- The importance of web testing;
- Basic contents of a test plan.

Speaking

Dialogue 1

A: If you have nothing special to point out, we'll have the sales contract made out tomorrow.

B: Good. By the way, the contract should be written in both English and Chinese, and both versions should be equally valid.

A: Of course. Each of us keeps one original and two copies.

B: Then I'll come along tomorrow to put my signature on it.

A: That's fine.

Dialogue 2

A: Shall we go over the other terms and conditions of the transaction to see if we agree on all the particulars?

B: All right. We have no objection to the stipulations about all the terms and conditions.

A: There is no question about that at our end.

B: Very good!

Dialogue 3

A: When would you like me to have the contract ready?

B: The earlier the better. I'm leaving next week.

A: How about next Monday afternoon at five? I'll have a copy of the contract sent to your hotel in the morning for you to look over.

B: Great!

Dialogue 4

A: We settled the terms for the contract two days ago. Now the draft contract is ready. Let's go over the terms and conditions of the contract. If you have any comments about them, do not hesitate to say so.

B: Okay. There is the last thing to make clear. How do we resolve the case where both parties hold different opinions on the standard of the goods?

A: Oh, suppose we have a dispute, we can resolve the case by submitting the dispute to arbitration by the Chinese International Trade Arbitration Commission.

B: All right, I'm glad our discussion has come to a successful conclusion.

Dialogue 5

A: Here is the contract. It contains basically all we have agreed upon during our negotiations.

B: To make sure no important items have been overlooked, let's check all the terms listed in the contract and see if there is anything not in conformity with the terms we have agreed on.

A: Okay. Let's start from the name of the commodity, specification, quantity, unit price…

B: Well, it looks good enough to me. You've done a good job.

Task: Create a short dialogue according to the following situation.

After reading the contract drafted by the secretary, you find one more provision should be added: the insurance premium should be born by the other party since the price is determined on a CFR basis.

Text Reading

Text A

Five Steps of Web Accessibility Testing

Testing the accessibility of a web page is an easy job. The necessary knowledge required is your familiarity with HTML and CSS, and your ability to predict the special difficulties faced by users. [1] The following steps are for your reference.

Step 1 It may come as a surprise that sometimes invalid code works as well. The code does not necessarily leave a visible bug. It may be logically correct but its writing style doesn't meet the standard. However, valid code tends to be easier to maintain as well as being compatible with many other technologies. [2] So first of all, in order to reduce possible problems the invalid code causes, validate your code.

Step 2　Automated accessibility testing is an important step in the overall process. When writing an article, rely on the spellchecker to catch your typos even though you know you still need to go through and check out the copy yourself to make sure you have written "Dave" and not "Cave", for instance. [3] Automated testing finds many problems which could easily be missed by reading the code.

Step 3　Hide your mouse and navigate your website using only your keyboard. Make sure every link, button, or any other functionality in the page is accessible via the keyboard. This is necessary because not all people use the mouse to browse websites.

Step 4　Try to carry out one or more of the tasks your website was built for. If it's an online store, find a product and buy it. If it's an informational site then find key information. Remember—this is the reason you built the site and it is the reason you are making it accessible.

Step 5　Web accessibility isn't just fulfilling a set of requirements or validating against predefined checkpoints. It also means quality design. So it's best to have your site checked over by an expert in screen reader and user interface design. They will have an intimate knowledge of their functions and use, the frustrations of poor website design and solutions which ease or eliminate those frustrations.

Web accessibility is something every company and organization needs to consider if they have or are planning to have a web presence. It's not only morally correct but increases your potential customer base. And this testing doesn't need to be difficult or expensive.

 **Notes:**

[1] The necessary knowledge required is your familiarity with HTML and CSS, and your ability to predict the special difficulties faced by users.

需要熟悉 HTML 和 CSS 的相关知识，以及能够预测使用者会遇到的特殊困难。

在本句中，required 过去分词做定语，修饰 knowledge；faced by users 过去分词做定语，修饰 difficulties。

[2] However, valid code tends to be easier to maintain as well as being compatible with many other technologies.

然而，有效代码更易于维护，更易于和许多其他技术兼容。

tend to 意为“倾向……”；be compatible with 意为“与……兼容”。

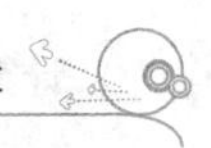

[3] When writing an article, rely on the spellchecker to catch your typos even though you know you still need to go through and check out the copy yourself to make sure you have written "Dave" and not "Cave", for instance.
写文章时，往往要依靠自动拼写检查发现打字错误，即使你还是要亲自浏览并检查确认写的是 Dave 而不是 Cave。

Vocabulary

accessibility [əkˌsesəˈbɪləti]	*n.*	易接近；可亲近；可得到
familiarity [fəˌmɪliˈærəti]	*n.*	熟悉，精通；亲密
invalid [ˈɪnvəlɪd]	*adj.*	无效的；作废的
code [kəʊd]	*n.*	代码，密码；编码；法典
visible [vɪzəbl]	*adj.*	明显的；看得见的；现有的
compatible [kəmˈpætɪb(ə)l]	*adj.*	兼容的；能共处的
potential [pəˈtenʃl]	*adj.*	潜在的；可能的
validate [ˈvælɪdeɪt]	*v.*	证实，验证；确认；使生效
automated [ˈɔːtəmeɪtɪd]	*adj.*	自动化的；机械化的
typo [ˈtaɪpəʊ]	*n.*	打印错误；打字（或排印）文稿的小错误
navigate [ˈnævɪgeɪt]	*v.*	驾驶，操纵；使通过；航行
functionality [fʌŋkʃəˈnæləti]	*n.*	功能；【数】函数性
via [ˈvaɪə]	*prep.*	通过；经由
browse [braʊz]	*v.*	浏览；漫不经心地看
predefined [ˌprɪdɪˈfaɪnd]	*adj.*	预先定义的
checkpoint [ˈtʃekpɔɪnt]	*n.*	检查站，关卡
interface [ˈɪntəfeɪs]	*n.*	界面，接口，交界面
intimate [ˈɪntɪmət]	*adj.*	亲密的；私人的
frustration [frʌˈstreɪʃn]	*n.*	挫折
eliminate [ɪˈlɪmɪneɪt]	*v.*	消除；排除

Phrases and Expressions

check out	检验；结账离开；通过考核
invalid code	无效代码

Exercises

Ⅰ. Answer the following questions briefly according to the text.

1. What's the necessary knowledge of testing the accessibility of a web page?
2. How can we reduce the possible problems caused by the invalid code?
3. What's the function of automated testing?

4. Why is keyboard testing necessary?

5. According to the passage, what's the importance of web accessibility?

Ⅱ. Fill in the blanks with words or phrases that match the meanings in the right column.

eliminate,	browse,	intimate,	familiarity,	invalid,	accessibility

1. ____________ a good knowledge of a particular subject or place
2. ____________ ineffective
3. ____________ having a close and friendly relationship
4. ____________ to remove or get rid of
5. ____________ the quality of being at hand when needed
6. ____________ reading superficially or at random

Ⅲ. Read the passage again and match each step with a proper title.

step 1	keyboard testing
step 2	validating the code
step 3	target audience testing
step 4	automated accessibility
step 5	screen reader testing

Ⅳ. Fill in each blank with the appropriate form of the words given in the brackets.

1. I want to upload data to the computer ______ (network) storage from my office computer.
2. The government has made an effort to make the Internet more ________ (access).
3. We regarded the contract as having become ________ (valid) because of lacking an essential element.
4. The channel will be open to ________ (navigate) as soon as the ice melts.
5. It was the first time they had attempted to talk ________ (intimate).
6. He had hoped to set a new world record, but ________ (frustrate) by bad weather.

Ⅴ. Translate the following paragraph into Chinese.

Web accessibility isn't just fulfilling a set of requirements or validating against predefined checkpoints. It also means quality design. So it's best to have your site checked over by an expert in screen reader and user interface design. They will have an intimate knowledge of their functions and use, the frustrations of poor website design and solutions which ease or eliminate those frustrations.

Text B

Introduction of Software Testing

Software testing is an activity that helps in finding out bugs in a software system under development, in order to provide a bug-free and reliable system to the customer.

Software manufacturers perform software testing to ensure the correctness, completeness and quality of their products. Their customers will be happy if the software works without any problem and will generate more profits for them.

A tester tries to find out all possible bugs in the system with the help of various inputs to it. Good testers know the technology well. They are tactful and know how to tell the developers, and persuade them to have the bugs fixed.

Apart from exposing bugs and confirming that the program meets the specifications, as a tester you need to create test cases, procedures, scripts and generate data. You execute test procedures and scripts, analyze standards and evaluate results of testing. You also:

- Speed up the development process by identifying bugs at an early stage;
- Reduce the organization's risk of legal liability;
- Assure successful launch of the product, save money, time and reputation of the company by discovering bugs and design flaws before failures occur in production.

A test plan is a document that describes the objectives, scope, approach, and focus of a software testing effort. The process of preparing a test plan is a useful way to think through the efforts needed to validate the acceptability of a software product. The completed document will help people outside the test group understand the "why" and "how" of product validation. It should be thorough enough to be useful but not so thorough that no one outside the test group can understand it.

At last, the testing will be completed according to the criteria specified in the test plan. This is when the success or failure of the system is decided based on the results. The test summary records this information. The summary provides the results of the designated testing activities and an evaluation of these results. Moreover, the summary provides an overall view of the testing and the quality of the software.

Vocabulary

bug [bʌg]	*n*	臭虫，小虫；故障；窃听器
software ['sɒf(t)weə]	*n*.	软件
reliable [rɪ'laɪəb(ə)l]	*adj*.	可靠的；可信赖的
generate ['dʒenəreɪt]	*v*.	使形成；使发生
profit ['prɒfɪt]	*n*.	利润；利益

input [ˈɪnpʊt]	*n.* & *v.*	输入
tactful [ˈtæktfʊl]	*adj.*	机智的；圆滑的；委婉的
expose [ɪkˈspəʊz]	*v.*	揭露；揭发；使曝光
specification [ˌspesɪfɪˈkeɪʃ(ə)n]	*n.*	规格；说明书；详细叙述
script [skrɪpt]	*n.*	脚本；手迹
execute [ˈeksɪkjuːt]	*v.*	实行；执行；处死
evaluate [ɪˈvæljueɪt]	*v.*	评估；估价
identify [aɪˈdentɪfaɪ]	*v.*	鉴定；识别；辨认出
liability [laɪəˈbɪlɪti]	*n*	可能性，责任；债务
launch [lɔːntʃ]	*v.*	发射；发起，发动
flaw [flɔː]	*n.*	瑕疵，缺点；裂缝，裂纹
approach [əˈprəʊtʃ]	*n.*	方法；途径；接近
criteria [kraɪˈtɪərɪə]	*n.*	（复数）标准；条件
specify [ˈspesɪfaɪ]	*v.*	指定；详细说明

Exercises

Ⅰ. Read the passage above and decide whether the following statements are true (T) or false (F).

( ) 1. The purpose of software testing is to find out bugs in a software system.

( ) 2. If the software works well, it will create more profits for the software manufacturers.

( ) 3. A tester has the abilities of finding out the possible bugs in the system and fixing them.

( ) 4. Good testers are necessarily good developers.

( ) 5. Reducing the organization's risk of legal liability is one of the tasks of a tester.

( ) 6. A test plan should be as thorough as possible so as to avoid misunderstanding.

( ) 7. The test summary provides a detailed view of the testing and the quality of the software.

Ⅱ. Fill in the blanks with the information given in the text.

1. Software testing is an activity that aims at providing a ________ system to the customer.

2. Software manufacturers perform software testing to ensure the ________, ________, and ________ of their products.

3. A good tester is ________ and knows how to persuade the developers to have ________ fixed.

4. A test plan is a ________ that describes the ________, ________, ________, and ________ of a software testing effort.
5. The test summary provides the results of the ________ and an ________ of these results.

Further Reading

Can You Trust Your Computer?

Who should your computer take its orders from? Most people think their computers should obey them, not obey someone else. With a plan they call "trusted computing", large media corporations (including the movie companies and record companies), together with computer companies such as Microsoft and Intel, are planning to make your computer obey them instead of you. (Microsoft's version of this scheme is called "Palladium".) Proprietary programs have included malicious features before, but this plan would make it universal.

Proprietary software means, fundamentally, that you don't control what it does; you can't study the source code, or change it. It's not surprising that clever businessmen find ways to use their control to put you at a disadvantage. Microsoft has done this several times: one version of Windows was designed to report to Microsoft all the software on your hard disk; a recent "security" upgrade in Windows Media Player required users to agree to new restrictions. But Microsoft is not alone: the KaZaa music-sharing software is designed so that KaZaa's business partner can rent out the use of your computer to their clients. These malicious features are often secret, but even once you know about them it is hard to remove them, since you don't have the source code.

In the past, these were isolated incidents. "Trusted computing" would make it pervasive. "Treacherous computing" is a more appropriate name, because the plan is designed to make sure your computer will systematically disobey you. In fact, it is designed to stop your computer from functioning as a general-purpose computer. Every operation may require explicit permission.

The technical idea underlying treacherous computing is that the computer includes a digital encryption and signature device, and the keys are kept secret from you. Proprietary programs will use this device to control which other programs you can run, which documents or data you can access, and what programs you can pass them to. These programs will continually download new authorization rules through the Internet, and impose those rules automatically on your work. If you don't allow your computer to obtain the new rules periodically from the Internet, some capabilities will automatically cease to function.

Today you can avoid being restricted by proprietary software by not using it. If you run GNU/Linux or another free operating system, and if you avoid installing proprietary applications on it, then you are in charge of what your computer does. If a free program has a malicious feature, other developers in the community will take it out, and you can use the corrected version. You can also run free application programs and tools on non-free operating systems; this falls short of fully giving you freedom, but many users do it.

Writing

Test Plan

测试计划是指用于描述所要完成的测试活动的范围、方法、资源和进度的文档，内容主要包括测试项、被测特性、测试人员和风险控制等。测试计划对测试全过程的组织、资源、原则等进行规定和约束，并制定测试全过程各个阶段的任务分配以及时间进度安排，在测试过程中可以不断修改与完善。

Table of Contents

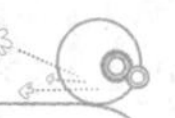

Task: The following is the table of contents of a test plan. Complete it by filling in the blanks according to the Chinese given in the brackets.

Test Plan

Table of Contents

Unit 5 Computer Marketing

Learning Objectives

In this unit you should get familiar with:

- Basic knowledge of marketing strategies;
- Various ways to do Internet marketing;
- Talking about promotion of products;
- Writing of order.

Speaking

Dialogue 1

A: Good morning, Lenovo Sales Department. What can I do for you?

B: Good morning. I'm Leo Wang from Oriental Import Co. We're interested in your enterprise desktops.

A: May I know which series do you like best?

B: I'm not sure. Could you explain to me in detail?

A: No problem.

Dialogue 2

A: We're going to update our desktops so as to suit our present work and your products impressed me a lot.

B: You're definitely right if you have interest in our products as they are developed just for users like your company, large, promising and eco-friendly.

A: But we think the price is a little high. We are wondering whether you can give us a discount.

B: Consider the high quality and all the features this series of products have, for instance, high speed, large storage, face recognition software, anti-bacteria keyboard, and it's really a big deal.

A: That's true.

Dialogue 3

A：I'm looking for a PC for myself.

B：What type of computer do you want?

A：I'd like to buy a laptop.

B：How much are you willing to pay for the computer?

A：I'm willing to pay up to 5,000 yuan for it.

B：What do you usually use the computer for?

A：I mostly use it to type documents and surf the Internet.

B：We have a special offer for students.

A：Oh，really? That's very good. I'll take it.

Dialogue 4

A：Hello，China Mobile. How can I help you?

B：Hello，I'm currently using a 10 yuan Text Message Plan. I'm considering switching to a new plan. Do you have some recommendations?

A：We have a full range of plans，from 20 yuan to 50 yuan. How many messages do you usually send per month?

B：About two or three hundred each month.

A：I would recommend the 30 yuan plan in that case. It offers 300 text messages.

B：OK. I'll switch to that plan then. Thank you!

A：You're welcome.

Dialogue 5

A：I'm considering choosing a plan for my cell phone.

B：We have a wide variety of plans for you to choose from. Our base plan is the 26 yuan plan，which covers 120 minutes' talk time.

A：That's not long enough. I travel a lot on business.

B：Then I'll recommend the 80 yuan plan. It's ideal for businessmen who travel a lot. It covers 900 minutes in total.

A：How much do you charge for additional minute?

B：20 fen per minute.

A：Sounds great. I'll go with this plan.

Text Reading

Text A

Business Marketing：Developing a Strategy

Marketing is a key part of business success. Developing a good business marketing

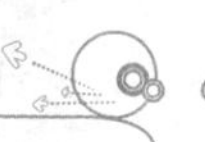

strategy[1] is what a company needs to rise above competitors. Careful research, proper introduction and good follow-up are very important to building a strong customer base.[2]

Research Your Market

Do you have a target market in mind? Get to know the customers you want. Find out what groups your products or services will attract the most. Find companies that need your products or services in the area. Carefully research your target market for an assessment of the things you may be able to offer. You can send out a short poll (no longer than 10 questions). It is a quick and easy way to find out what they need.

Examining the competition closely will help to find areas of demands your company can fill in.[3] Figure out what special products or services your company offers that others do not. Studies show that specialty companies do better than those who try to do it all.

Introduce Your Business

Start out with a simple handout. Any piece of marketing should include these basic elements: who you are, what you offer, where you are located, and how to contact you.

Try to interest people by emphasizing those unique specialties that only your business offers in your advertising. The thing to focus on is one or two particular needs they have that you can take care of for them.

Other things you might include in an advertisement are: grand opening date, special sales, top-selling products, number of customers to date, and company slogan/logo.

Follow-Up

One of the aspects that people pay less attention to of marketing is follow-up. It is always good to have new customers, but to truly be successful, you need at least some of those people to come again. There are some very simple things that you can do to inspire repeat business.

If you get phone calls or e-mails, respond to them. Give a call back or a reply message to thank them again for inquiring. Offering coupons, discounts or a free item (e.g. a pen with company logo) is also a way to thank and encourage them to come again.

If you follow these steps, you will surely see your business take off. Once you find out what works for you, just do it again and again, and you will soon have a strong customer base.

Notes:

[1] marketing strategy

营销策略是企业以顾客需要为出发点，有计划地组织各项经营活动，为顾客提供满意的商品和服务并实现企业目标的过程。

[2] Developing a good business marketing strategy is what a company needs to rise above competitors. Careful research, proper introduction and good follow-up are very important to building a strong customer base.
企业要想在竞争对手中脱颖而出就需要开发一套好的营销策略。仔细的调研、适当的宣传和及时的跟进对建立一个牢固的客户群来说都是十分重要的。
"what" 在句中引导表语从句；"rise above competitors" 在本句中可译为"在竞争对手中脱颖而出"。

[3] Examining the competition closely will help to find areas of demands your company can fill in.
仔细研究竞争情况能够帮你找到一些你的公司可以填补的需求空白。

Vocabulary

strategy ['strætɪdʒɪ]	*n*.	战略，策略
specialty ['speʃ(ə)lti]	*n*.	专业，专长；特产；特性
poll [pəul]	*n*.	投票；民意测验；投票数
grand [grænd]	*adj*.	宏伟的；豪华的；极重要的
slogan ['sləug(ə)n]	*n*.	标语；呐喊声
assessment [ə'sesmənt]	*n*.	评定；估价
logo [ləugəu]	*n*.	商标，徽标；标志语
inspire [ɪn:spəɪə]	*v*.	激发；鼓舞；启示；产生
coupon [ku:pɒn]	*n*.	息票；赠券；联票
follow-up ['fɒləu ʌp]	*n*.	随访；跟进；后续行动
item ['aɪtəm]	*n*.	条款，项目；一件商品
competitor [kəm'petɪtə]	*n*.	竞争者，对手
emphasize ['emfəsaɪz]	*v*.	强调，着重

Phrases and Expressions

customer base	客户群
figure out	解决；算出；想出
start out	出发；着手进行
take off	起飞；脱下；突然成功

Exercises

Ⅰ. Answer the following questions briefly according to the text.

1. What are important in building a strong customer base?
2. How can people do the market research?

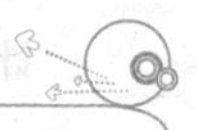

3. What's the purpose of examining the competition?
4. What basic information should be included in a simple handout?
5. What other things you may include in an advertisement?
6. What can be done to inspire repeat business?

Ⅱ. **Match the words with their explanations in the right column.**

1. assessment	come to understand
2. respond	give particular importance to sth.
3. take off	get down to
4. figure out	a person that is competing with another
5. emphasize	a process to make a judgment about a person
6. start out	achieve sudden growth or success
7. competitor	reply

Ⅲ. **Complete the sentences with the words or phrases given in the table. Change the form if necessary.**

follow-up,	poll,	competitor,	specialty,	handout,
slogan,	logo,	coupon,	assessment,	emphasize

1. Before you buy something online, do a ________ search and potentially it can save yourself a lot of money.
2. Home-made ice-cream is one of our ________.
3. It is a long term illness and regular ________ appointments are required.
4. A ________ is a design that is used by an organization for its letterhead, advertising material, and signs as an emblem by which the organization can easily be recognized.
5. My teacher always ________ the importance of grammar in English learning.
6. Please read the ________ carefully to get more information about the conference.
7. I totally agree with your ________ of the situation.
8. Last week, we carried out a ________ and found out that the support for the headmaster was strong.
9. We sell twice as many laptops as our ________.
10. He finished his speech with the same ________.

Ⅳ. **Choose the best words or phrases to complete the following passage.**

Computer hackers have now got their hands on mobile phones. A phone virus can ___1___ your phone do things you have no control over, computer security experts ___2___. It might ___3___ the White House or the police, or forward your personal address book to a marketing company. Similar viruses have already made mobile phone owners ____4____ in Japan and Europe.

Mobiles are now able to surf the net, send e-mails and ____5____ software, so they are an easy ____6____ for the same hackers who have sent viruses to computers over the last decade. If the phone is connected to the ____7____, it can be used to transmit threats and ____8____ targets, just as any computer can. In Japan, if you opened a certain e-mail message ____9____ your mobile, it would cause the phone to repeatedly ____10____ the national emergency number. So phone operators had to ____11____ emergency calls until the bug was ____12____. In Europe, mobile's short message service, ____13____ SMS, has been used to send codes that could damage phones.

Mobile users can ____14____ viruses, of course, by sticking to their traditional phones ____15____ Web links, some experts said.

(　　) 1. A. get	B. force	C. make	D. damage
(　　) 2. A. speak	B. talk	C. tell	D. say
(　　) 3. A. lead	B. cause	C. control	D. call
(　　) 4. A. interested	B. angry	C. excited	D. satisfied
(　　) 5. A. make	B. destroy	C. download	D. develop
(　　) 6. A. job	B. task	C. mission	D. target
(　　) 7. A. computer	B. television	C. Internet	D. radio
(　　) 8. A. strike	B. visit	C. inquire	D. attack
(　　) 9. A. in	B. by	C. on	D. with
(　　) 10. A. send	B. dial	C. count	D. press
(　　) 11. A. cancel	B. ban	C. stop	D. prevent
(　　) 12. A. removed	B. cleaned	C. called	D. cleared
(　　) 13. A. and	B. nor	C. or	D. but
(　　) 14. A. stop	B. avoid	C. kill	D. find
(　　) 15. A. beyond	B. with	C. over	D. without

Ⅴ. Translate the following paragraph into Chinese.

The tireless head of HP's Personal Computer Department is winning over competitors Dell and Lenovo. He has helped the company become the top PC maker in the world and has increased profits at the same time. Bradley's strategy has depended on creating fresh designs and eye-catching marketing campaigns. Hip-hop star Jay Zhou and tennis star Williams talked about using HP PCs to watch movies without first waiting for Windows to launch as well as to look at photos and edit music. The TV ad ended with a slogan: "The computer is personal again."

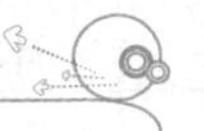

Text B

Heaven for Taobao. com's Shop Owners

Could you believe that 70 percent of the rooms of a business building in Hangzhou belong to hundreds of online-shop owners for China's biggest auction website Taobao. com? The 20-floor building called Sijixingzuo, which means "Constellations of the Four Seasons", is also nicknamed Taobao building as several hundred successful online shop owners of Taobao. com have rented most of the apartment inside the building. It locates just opposite to Hangzhou's biggest fashion and clothing market—the Evergreen Market in eastern downtown Hangzhou.

The newspaper reported that these brisk businesses are now responsible for sending out over 10,000 packages per day. They are not only bringing over to themselves huge amount of profit, but also adding vitality to the local economy and the job market.

Taobao, which means hunting for valuable things in Chinese, offers free online auction services. Most of the business that are operating in Taobao building are run by young people in their twenties trading clothes, accessories and shoes etc. The business model for Taobao aims to cut business cost, resulting in the win-win situation between manufacturers and consumers. Consumers can often get a great bargain or find a hard-to-find object. At present, the major consumer group of Taobao is aged between 18 and 35. They pursue fashion and have relatively higher consuming capability.

Zhang Nan is one of the shop owners in the Taobao building. He said he used to run his web business at home, however, he was constantly pressed for time because the wholesale market where he got his supplies was too far away. Then he heard about the Taobao building, where most of the tenants are young people and where it is very convenient to organize fresh suppliers. After checking out the location Zhang decided to open a store here. Zhang says many of the shop owners are graduates from designing schools. They make the designs and get a plant in the suburb of Hangzhou to produce their own goods. This cooperation helps keep their costs down. Taobao building is a very good place for college students to start up business, says Zhang. The rent is acceptable, and they are allowed to share the same storefront.

Vocabulary

auction [ɔːkʃ(ə)n]	*n.& v.*	拍卖，竞卖
constellation [ˌkɒnstəˈleɪʃ(ə)n]	*n.*	星座，一系列（相关事物）
nickname [ˈnɪkneɪm]	*n.& v.*	绰号，昵称；给……起绰号
locate [ləʊˈkeɪt]	*v.*	坐落于；设置

downtown [ˈdaʊntaʊn]	*adj*. & *adv*.	商业区的；市中心的；在商业区
brisk [brɪsk]	*adj*.	轻快的，活泼的，繁忙的
profit [prɒfɪt]	*n*.	利润
accessory [əkˈses(ə)rɪ]	*n*.	配饰，配件
manufacturer [ˌmænjuˈfæktʃərə]	*n*.	制造商，厂商
bargain [ˈbɑːgən]	*n*. & *v*.	便宜货；讨价还价
consumer [kənˈsjuːmə]	*n*.	消费者，用户，顾客
pursue [pəˈsjuː]	*v*.	从事；追赶；纠缠
capability [ˌkeɪpəˈbɪlətɪ]	*n*.	才能；能力；性能；容量
constantly [ˌkɒnst(ə)ntli]	*adv*.	经常地，不断地
pressed [prest]	*adj*.	紧缺的，缺少的
wholesale [ˈhəʊlseɪl]	*n*. & *v*.	批发
tenant [ˈtenənt]	*n*.	房客；承租人
storefront [ˈstɔːfrʌnt]	*n*.	店面，铺面
vitality [vaɪˈtælɪtɪ]	*n*.	精神，活力，力量，热情

Phrases and Expressions

opposite to	在……对面
send out	发送；放出；派遣
bring over	带来；使……相信；说服
check out	检验；核对；结账离开
start up	开始；启动

Exercises

Ⅰ. Choose the best answers according to the passage.

1. There are ________ floors in the Taobao building.

   A. sixteen　　B. twelve　　C. twenty　　D. thirty

2. The building is located ________ the Evergreen Market in eastern downtown Hangzhou.

   A. in front of　　B. opposite to　　C. inside　　D. next to

3. The products sold in Taobao building do not include ________.

   A. houses　　B. shoes　　C. accessories　　D. clothes

4. The major consumer group includes ________.

   A. retired people　　B. children　　C. young people　　D. old woman

5. Mr. Zhang chooses Taobao building to run his web business NOT because ________.

   A. it is very convenient to organize fresh suppliers

B. the rent is acceptable

C. the tenants are allowed to share the same storefront

D. it is far away from his home

Ⅱ. Translating.

A. Translate the following sentences into English.

1. 他们决定明年搬到上海的郊区去。(suburb)
2. 我现在不能停下来——我时间有点紧。(pressed)
3. 她的商店现在每年盈利100多万美元。(profit)
4. 莉莉为婚礼准备了一件带配套小饰物的礼服。(accessory)
5. 这个国家有能力制造核武器。(capability)

B. Translate the following sentences into Chinese.

1. I have rented a house and paid the rent.
2. He took a seat opposite to me, smiling at me.
3. Even though she is in her 80s, she's still full of vitality.
4. Jim is constantly changing her mind.
5. She hopes to pursue a career in film-making in the future.

Further Reading

Internet Marketing

When we think of traditional marketing methods we think immediately of ads in the press (both local and national), page space in a variety of publications. And, naturally, as business people with budgets to consider, we think of the costs of such ventures and then consider some of the most effective forms of marketing, for example, the Internet marketing. The Internet has brought many unique benefits to marketing including low costs in distributing information and a media to global audience.

Most businesses who have embraced a heavy traditional marketing campaign in the past will know that keeping track of where initial contacts come from is not an easy task. Of course you can always inquire as to how a new client heard about your company, but that's time consuming. Internet marketing makes tracking easy. You can keep tabs on where your visitors come from, store relevant information about their spending, their behaviors, and their demographics. All of this better enables your company to make the most of Internet marketing to learn more about your existing and potential clients. More importantly, you'll very quickly get a clear picture of what works for your company online.

Times are changing and your prospective clients are far more likely to 'Google' their needs as a first port of call than go anywhere else for information. A difference between traditional methods and Internet marketing mainly lies in cost. The costs are certainly lower with the former and the effects of online promotions are more far-reaching and have a higher impact. However, there are also some limitations of Internet marketing. It does not allow the customers to touch, smell, taste or try on tangible goods before making an online purchase.

Writing

Order

订单是为了要求供应具体数量的货物而提出的一种要求。它是对报盘或询盘后发出报价而促成的结果。订单可以用信或印制好的订单、传真或电子邮件来发送。卖方则用印制好的确认书来回复。

订单的主要特点是正确和清楚，至少应包括以下内容。

1）详细的说明、数量、价格以及货号等。

2）说明包装方式、目的港以及装运期。

3）确认在初期洽谈时所同意的支付条件。

4）收货人的详细信息。

Sample 1

Initial Order

Dear Sir/Madam,

Thank you for your letter of March 20th. We are pleased to find that your products are of excellent quality.

As a trial order, we are delighted to give you a small order for 100 sets of your Air Conditioners BC-163. Please note that the goods are to be supplied in accordance with your samples.

Enclosed you will find Order Sheet No. 137. As we are in an urgent need of the commodity for the coming season of sales, please fax your acceptance. Upon receipt of which, we will open an irrevocable L/C through the Bank of China.

Yours faithfully,

Zhang Lin

Sample 2

Order Sheet

Dear Sir/Madam,

We have the pleasure of placing the following order with you on the terms and conditions as set forth.

Article: Haier Brand Air Conditioners

Description: Exactly the same as Sample 103

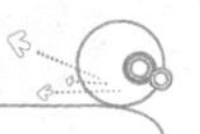

Quantity：100 sets

Unit Price：US ＄805 per set CIF Los Angeles Ca. USA

Amount：US ＄80，500

Shipment：During March

Packing：One set in a cardboard carton，ten cartons in a wooden case

Shipping Marks：(omitted)

Insurance：AAR for full invoice amount plus 10%

Payment terms：Draft at 30 days' sight under an irrevocable L/C

Remarks：Certificate of quality and shipment samples to be sent by air mail prior to shipment.

Yours faithfully，

Wu Yue

Task：Translate the following sentences into English with the help of the words given in the bracket.

1. 我们合同中对付款条件有详细规定。(specify)
2. 我们相信首次订货会导致将来贸易的进一步发展。(result in)
3. 由于我方急需，如你方尽早安排发货我们将不胜感激。(dispatch)
4. 我们会确保货物准时发运至你方港口。(prompt dispatch)
5. 由于你方的报价有竞争力而且能让我们接受，我们考虑大量购买。（quotation，large quantities)
6. 由于订单太多，我们不能接受新订单，但一旦有新货，我们将立即与你方联系。(fully committed)

Unit 6 Computer Distribution

Learning Objectives

In this unit you should get familiar with:

- Different kinds of packaging;
- The main functions of packaging;
- Importance of packaging;
- Basic knowledge of instructions.

Speaking

Dialogue 1

A: Would you tell me how you plan to pack our order?

B: For the monitors, we use a polybag for each set and then pack it in a box. Our packing appeals strongly to consumers.

A: How can you protect your products from pressing and shocking?

B: Well, each case is lined with foam plastics to prevent pressing and shocking.

A: I'm afraid the boxes are not strong enough for sea transportation.

B: Please don't worry about that. Each box is lined with plastic sheets from inside so as not to be spoiled by dampness or rain.

Dialogue 2

A: How are you going to pack my scanners?

B: We'll pack them as follows: one set per box, 6 boxes to a carton. What do you think?

A: As you know, packing is very important to the sales promotion. How are you going to interest our customers with packing?

B: The box will be attractive, durable and easy to handle. I'm sure the colorful photos and the descriptions printed on the boxes will be eye-catching.

A: It sounds nice. Please don't forget to print the shipping marks on the outer packing.

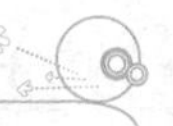

B: No problem. We'll try our best to pack the goods to your satisfaction.

Dialogue 3

A: How are you going to pack these computers?

B: I take some packaging samples with me. Please have a look and choose the one you're satisfied.

A: This one looks nice and fashionable.

B: It's also my favorite.

Dialogue 4

A: I think we should discuss further about the terms of packaging.

B: What's your proposal?

A: You know the chemical drugs should be promoted carefully against moisture. So they are not safe enough to be packed in paper cartons.

B: Then what is your suggestion?

A: They should be packed in plastic-lined water-proof cartons.

B: How about polythene packaging?

A: OK. That's great.

Dialogue 5

A: The next thing I'd like to bring up for discussion is packing.

B: We'd like to hear what you say concerning the matter of packing.

A: You know the crux of packing lies in protecting the goods from moisture. So shirts are to be packed in plastic-lined water-proof cartons.

B: We use a polythene wrapper for each shirt and the shirts are packed in cardboard boxes of forties each.

A: It's fine. And we want the label to be fixed on the polythene wrapped.

B: OK. We'll do that.

Useful Sentences

1. We have definite ways of packing...
2. As to silk blouses, we use... all ready for window display.
3. I hope the packing will be...
4. We have here a sample packing...
5. Do you mind if I give you a little suggestion?
6. Your packing needs improvement. I mean...
7. Your suggestion on packing is welcome as well.
8. What's your outer packing?
9. Cardboard boxes are light to handle and less expensive while wooden cases are clumsy and cost more.

Text Reading

Text A

The Functions of Packaging

Packaging is the technique of preparing goods for distribution. In recent years, the significance of packing has been increasingly recognized, and today the widespread use of packing is truly a major competitive force in the struggle for markets. The linking of packaging with integrated logistics is no more evident than with transportation. [1]

Packaging can be divided into industrial packaging and consumer packaging. Generally speaking, consumer packaging, which mainly aims at containing the goods, promoting the sale of it and facilitating use of it, is of little value to logistics operation. But industrial packaging has a significant impact on the cost and productivity of logistics. Industrial packaging should perform the following functions to meet integrated logistics requirements.

First, it should protect the goods from damage during handling, storing and transportation. Damage caused by vibration, impact, puncture or compression can happen whenever a package is being transported. Hence, package design and material must combine to achieve the desired level of protection without incurring the expense of over protection. It is possible to design a package that has the correct material content but does not provide a necessary protection. Arriving at a satisfactory solution involves defining the degree of allowable damage in terms of expected overall conditions (because in most cases, the cost of absolute protection will be prohibitive) and then isolating a combination of design and material capable of meeting those specifications. [2]

Second, it should promote logistical efficiency. Packaging affects not only marketing and production but also integrated logistics activities. For example, the size, shape and type of packaging material influence the type and amount of material handling equipment as well as how goods are stored in the warehouse. Likewise, package size and shape affects loading, unloading, and the transporting of a product. The easier it is to handle a product, the lower the transportation rate. Hence, if the package is designed for efficient logistical processing, overall system performance will benefit.

The third important logistical packaging function is communication or information transfer. To identify package contents for receiving, order selection and shipment verification, etc., is the most obvious communication role of packaging. Typical information includes manufacturer, product, container type, count, and Universal Product Code (UPC) number. Ease of packaging tracking is also important. Effective internal operation and a growing number of customers require that product be tracked as

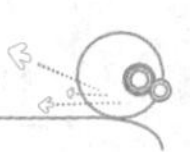

it moves through the logistics channel. This can be realized by the extensive use of Radio Frequency Identification (RFI), a computer chip embedded in the package, container, or vehicle to allow the container and contents to be scanned and verified as it passes checkpoints in the distribution facility and transportation gateway. [3] The final communication role of logistics packaging is to provide instructions as to how to handle the cargo and how to prevent possible damage. For instance, if the product is potentially dangerous, such as fireworks and table tennis balls, the packaging or accompanying material should provide instructions for avoiding moisture, vibration and heating, etc., as the case may be.

In addition to product protection, packages should be easy to handle, convenient to store, readily identifiable, secure and of a shape that makes best use of space. [4] Thanks to packaging, it's possible for products to be available anytime anywhere that gives a consumer a great freedom of choice.

Notes:

[1] The linking of packaging with integrated logistics is no more evident than with transportation.
包装与一体化物流的关系最明显体现在运输上。
The structure "no more evident than" has similar meaning to "the most evident" (最明显的).

[2] Arriving at a satisfactory solution involves defining the degree of allowable damage in terms of expected overall conditions (because in most cases, the cost of absolute protection will be prohibitive) and then isolating a combination of design and material capable of meeting those specifications.
要满意地解决包装问题，就需要根据预期的总体状况，判断货物所允许发生的货损程度（因为在大多数情况下，不会为了完全避免货损而使成本过高），然后找出一种包装，供设计和选材符合这些要求。

[3] This can be realized by the extensive use of Radio Frequency Identification (RFI), a computer chip embedded in the package, container, or vehicle to allow the container and contents to be scanned and verified as it passes checkpoints in the distribution facility and transportation gateway.
射频识别系统的广泛应用，使这一切得以实现。该系统在包装品、容器或运输工具中内置一块计算机芯片，便可以使其在经过配货站和运输出入口的各检查点时，包装容器和包装内容能够被扫描并得到查证确认。

[4] In addition to product protection, packages should be easy to handle, convenient to store, readily identifiable, secure and of a shape that makes best use of space.
除去商品保护的功能，包装还应该易于处理，方便存放，容易识别，安全，且拥有把空间最大化利用的外形。

Vocabulary

promote [prə'məʊt]	*v*.	促进；提升；推销；发扬
facilitate [fə'sɪlɪteɪt]	*v*.	促进；帮助；使容易
logistics [lə'dʒɪstɪks]	*n*.	物流
significant [sig'nɪfɪk(ə)nt]	*adj*.	重大的；有意义的；意味深长的
productivity [prɒdʌk'tɪvɪtɪ]	*n*.	生产力；生产率；生产能力
integrate ['ɪntɪgreɪt]	*v*.	使……完整；使……成整体
vibration [vaɪ'breɪʃ(ə)n]	*n*.	震动，颤动
puncture ['pʌŋ(k)tʃə]	*n*. & *v*.	刺穿
compression [kəm'preʃ(ə)n]	*n*.	压缩，浓缩；压榨，压迫
hence [hens]	*adv*.	因此；今后
incur [ɪn'kɜː]	*v*.	招致，蒙受，遭遇
prohibitive [prə'hɪbɪtɪv]	*adj*.	禁止的，禁止性的，抑制的
isolate ['aɪsəleɪt]	*v*.	使隔离；使孤立；使绝缘
warehouse ['weəhaus]	*n*.	仓库；货栈；大商店
transfer [træns'fɜː]	*v*.	转让；转移；传递；过户
identify [aɪ'dentɪfaɪ]	*v*.	确定；鉴定；识别，辨认出
verification [ˌverɪfɪ'keɪʃ(ə)n]	*n*.	确认，核实；查证
embed [ɪm'bed]	*v*.	栽种；使嵌入
vehicle ['viːəkl]	*n*.	车辆；工具；交通工具；运载工具
verify [verɪfaɪ]	*v*.	核实；查证
checkpoint ['tʃekpɔɪnt]	*n*.	检查站；关卡
gateway ['geɪtweɪ]	*n*.	网关；通道；途径
cargo ['kɑːgəu]	*n*.	货物
fireworks ['faɪəwɜːks]	*n*.	烟火
moisture ['mɔɪstʃə]	*n*.	潮湿；湿气
readily ['redɪli]	*adv*.	容易地；乐意地；无困难地
identifiable [aɪdentɪ'faɪəb(ə)l]	*adj*.	可辨认的
potentially [pə'tenʃəli]	*adv*.	可能地；潜在地

Exercises

Ⅰ. Answer the following questions briefly according to the text.

1. What're the aims of consumer packaging?
2. What're the functions of industrial packaging?
3. Which factors may cause the damage when a package is being transported?
4. How does packaging influence integrated logistics activities?

5. What is the most obvious communication role of packaging?

6. What is the final communication role of logistics packaging?

Ⅱ. **Fill in the blanks with the words or phrases given below. Change the form if necessary.**

motivate,	consumer,	package,	shipping,	potential,	information
essential,	purchase,	transport,	distinctive,	comply with,	protect

Today, good ___1___ design is regarded as an ___2___ part of successful business practice. Since many ___3___ customers first notice a new product after it has arrived on the shelves of a store, it is vital that the packaging provide ___4___ with the ___5___ they need and ___6___ them to make a ___7___. But packaging decisions involve a number of tradeoffs. While making a product visible and ___8___ may be the top priority, for example, businesses must also ___9___ a variety of laws regarding product labeling and safety. ___10___ products during ___11___ is important, but businesses also need to keep their ___12___ costs as low as possible.

Ⅲ. **Match the indicative marks and the warning marks with their Chinese meanings.**

Handle With Care　　最高堆层
This Side Up　　危险物品
Inflammable　　此端向上
Use No Hook　　易燃物品
Sling Here　　由此吊起
Maximum Stack　　易腐物品
Fragile　　小心轻放
Dangerous Goods　　不得用钓钩
Perishable Goods　　冷藏
Keep Cool　　易碎物品

Ⅳ. **Read the text again and fill in the blanks with the information given in the text.**

1. Today the widespread use of packing is truly a ________ in the struggle for markets.

2. Packaging can be divided into ________ and ________.

3. Industrial packaging has a significant impact on the cost and ________.

4. Packaging should protect the goods from damage during ________, ________ and ________.

5. The easier it is to handle a product, the lower ________.

6. It's possible for packaging products to be available anytime anywhere that gives a consumer ________.

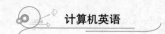

Ⅴ. **Translate the following paragraph into Chinese.**

Packing is one of the most important ways to realize the value of commodities. It can protect and prettify commodities and form an important process in the storage, transportation and sales of commodities. Packing mark is the mark to facilitate the cargo, transfer, stock or the mark to identify the goods and avoid damage. Packing mark includes shipping mark, indicative mark, warning mark and so on.

Text B

Factors Influencing Types of Cargo Packaging

Packaging in the physical environment is subject to moisture, temperature extremes, mechanical shocks and vibration. No matter what environmental conditions are encountered, the package is expected to protect the product, while keeping it in the condition intended for use until the product is delivered to the ultimate consumer. There are several factors to be considered to decide the type of packaging.

(1) Value of goods. The high-value consignment usually attracts more extensive packing than low-value merchandise. If packing is inadequate, problems could be experienced in carriers' liability, acceptance and adequate cargo insurance coverage. Moreover, high-value consignments, such as a valuable painting, require adequate security and likewise attract higher freight rates.

(2) Nature of transit. That is to say, the type and length of the transit. One must consider the method of shipment; it may be LCL or FCL. The more handling goods must endure, the stouter the packaging required. The greater the degree of over stowage that the goods must endure, then the stouter the packaging needed. Certain forms of transport, particularly air and ISO containerization, usually require less extensive packaging. This is a strong marketing feature of their service. Air transport particularly encourages palletized consignments with cargo strapped throughout the transit.

(3) Nature of the cargo. This concerns the characteristics of the goods concerned and their susceptibility to various loss or damage. It is important to bear in mind that packaging offers protection against pilferage, as well as damage. This factor together with item is the two main factors which determine the type of packing for an individual consignment. Cargo shipped in bulk requires little or no packing, while general merchandise needs adequate packing.

(4) Variation in temperature during the course of the transit. Temperature variation can be quite extensive during transit and packaging must take account of this to permit

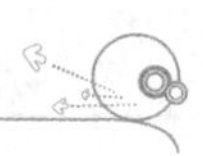

the cargo to breathe and avoid excessive condensation or sweating. Again, advice should be sought from the airline or ship-owner when necessary.

Packing, therefore, is not only designed as a form of protection to reduce the risk of goods being damaged in transit, but also to prevent pilferage and aid marketing. It is, of course, essential to see not only that the right type packing is provided, but also that the correct quality and form of container is used.

Vocabulary

extreme [ɪk'striːm]	*n.*	极端；末端；极端的事物
encounter [ɪn'kaʊntə]	*v.*	偶然碰到；遇到；遭遇
ultimate ['ʌltɪmət]	*adj.*	最终的；极限的
consignment [kən'saɪnm(ə)nt]	*n.*	委托；货品；货物
extensive [ɪk'stensɪv]	*adj.*	广泛的；大量的；广阔的
merchandise ['mɜːtʃ(ə)ndaɪs]	*n.*	商品；货物
inadequate [ɪn'ædɪkwət]	*adj.*	不充分的；不恰当的
insurance [ɪn'ʃʊər(ə)ns]	*n.*	保险；保险费
freight [freɪt]	*n.*	货运；运费
transit ['trænsɪt]	*v.*	运输；经过
endure [ɪn'djʊə]	*v.*	忍耐；容忍；持续
overstowage [ˌəʊvə'stəʊɪdʒ]	*n.*	货物倒装
stout [staʊt]	*adj.*	结实的；矮胖的；激烈的
containerization [kənˌtenərɪ'zeʃən]	*n.*	集装箱化；集装箱运输
palletize ['pæləˌtaɪz]	*v.*	用托盘装运
strap [stræp]	*n.* & *v.*	用带捆绑
susceptibility [səˌseptɪ'bɪlɪti]	*n.*	易受感染；易受影响；敏感性
pilferage ['pɪlfərɪdʒ]	*n.*	盗窃
variation [veərɪ'eɪʃ(ə)n]	*n.*	变化；变动；变异
excessive [ɪk'sesɪv]	*adj.*	过多的；极度的；过分的
condensation [kɒnden'seɪʃ(ə)n]	*n.*	冷凝；凝结；压缩
sweat [swet]	*n.* & *v.*	汗珠；出汗；冒汗
container [kən'teɪnə]	*n.*	集装箱；容器

Phrases and Expressions

LCL：less than container load	拼箱货
FCL：full container load	整箱货
ISO：International Organization for Standardization	国际标准化组织

Exercises

Ⅰ. **Choose the best answers according to the passage.**

1. The factors influencing the packaging in the physical environment do NOT include ________.

   A. moisture　　B. mechanical shocks

   C. temperature extremes　　D. quantities of the products

2. The function of packaging is to keep the product ________.

   A. as beautiful as possible　　B. clean

   C. in the condition intended for use　　D. under control

3. With inadequate packing, problems may arise in ________.

   A. carriers' liability　　B. acceptance

   C. adequate cargo insurance coverage　　D. all of the above

4. Which one of the following transports requires stricter packaging?

   A. Air transport.　　B. Railway transport.

   C. Highway transport.　　D. Waterway transport.

5. Packaging must take account of ________ to permit the cargo to breathe and avoid excessive condensation or sweating.

   A. pilferage　　B. temperature variation during transit

   C. damage　　D. marketing

Ⅱ. **Fill in each blank with the appropriate form of the word given in the brackets.**

1. She falls into depression and ________ (ultimate) commits suicide.
2. We have ________ (encounter) so many problems in building the bridge.
3. In consideration of your ________ (extensive) experience in the field, we are glad to appoint you as our agent.
4. I'm sure Susan is the best candidate for the job because her knowledge of English is ________ (inadequate) for the intercultural communication.
5. This company has admitted ________ (liability) for the accident.
6. The prisoners had to ________ (endure) months of hunger.
7. An ________ (excessive) of fat in one's diet can lead to heart disease.
8. We are not a bit afraid of bleeding and ________ (sweat).

Further Reading

Distribution

With the rapid development of the society, companies adopt a series of methods to sell their products in response to enormous changes in market in order to achieve a

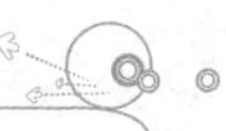

splendid success. Generally, they use direct and indirect distribution. What kind of marketing strategy will contribute to further prosperity of a nation among so many distributions?

As the word implies, direct distribution method refer to the selling of products or services to the consumer directly without the use of intermediaries. Indirect distribution for manufactured products includes the use of one or more intermediaries such as manufacturers, wholesalers, retailers, consumers. It is divided into intensive distribution, selective distribution and exclusive distribution.

Different advantages and disadvantages can be found in direct distribution and indirect distribution method. The direct distribution requires low cost and it makes products cheaper. Of course the company can make supernormal profit. On the contrary, in indirect distribution, the risk was shared, so does the profit. It leads to a higher price for customers. It seems that the first method is better for us due to the superficial lower price. But in fact indirect distribution creates a lot of jobs for our country. Therefore, it is not only good to Chinese large population but also national steady and booming.

Lenovo Group is a Chinese company which is famous for its excellent electronic products especially when it comes to its extraordinarily functional computer.

As we all know, Lenovo, as a national PC brand, presents Chinese wisdom and intelligence to the world. Lenovo Group uses depth distribution to improve the sales volume. There are always several general agents in every province. Except the general agents, they also employ a lot of distributors, secondary distributors and so on. Through so many intermediaries, the price of the Lenovo computers maybe is much higher when compared with being sold in the direct distribution method. Superficially, it is not do any good to the considerable customers. But because of it, Lenovo Group helps more Chinese own an ideal job.

Writing

Instructions

产品说明书是关于产品用途、规格和使用方法等的文字说明。其主要功能是向消费者提供具体的商品特点和使用信息，起到了指导消费的作用。产品说明书的语言应浅显确切、简洁明了、逻辑清楚，主要特点是大量使用省略句、祈使句和被动语态。一般来说，一则完整的产品说明书包括以下几个部分：

（1）产品的概况（如名称、产地、规格、制作方法、主要成分、特征、功能等）。

（2）产品的性能、指标和规格。

（3）产品的安装、操作和使用方法。

（4）保养和维修方法。

（5）附件、备件及其他需要说明的内容。

Sample 1

Zipper Lock

Usage Instructions

The lock is set at the manufacturer to open at 000. You can keep it as your combination，or set a new one as following steps：

1）Push the button in the direction of the arrow and hold it，until completing the next step.

2）Turn the dials to your desired personal combination.

3）Release the button and rotate the dials so that your three secret numbers can't be seen.

4）Now your new personal code is set.

5）If you want to change to a new combination，repeat the steps.

Sample 2

Kenwood Dynamic Microphone MC-550

Instruction Manual

*Operating Instructions

Insert the microphone plug into the microphone terminal.

Flip the microphone switch to the "ON" position and adjust the volume with the volume control knob on the amplifier.

When handling the microphone to somebody else or when finish using，flip the microphone switch to the "OFF" position.

*Handling Precautions

If the microphone head is covered by hand or the microphone is approached to the speaker，a howling sound may be generated. This phenomenon of howling is caused by the microphone picking up the sound output from the speaker. To prevent this，first decrease the volume，then place the microphone so that it is not pointed to the speaker and that there is a sufficient distance between the microphone and the speaker.

The microphone is sensitive equipment. Do not drop，hit it or apply strong shock to it.

Do not store the microphone in a place with high temperature or humidity.

*Pointer for Proper Use of Microphone

The optimum distance between the microphone and the mouth is from 5 to 10 centimeters. If the microphone is too close to the mouth，the sound may be unclear with too much enhanced base（proximity effect）or may be uncomfortable to ears with pop

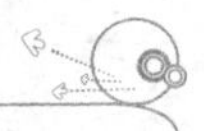

noise generated every time when the singer breathes in and out.

Task: Complete the instructions below by putting the Chinese in the brackets into English.

1. Before transporting the scanner, please move the protection switch to ___________（关闭位置）.
2. Place a ___________（彩色图片）facing down on the glass plate of the scanner.
3. You can double click on the thumbnail to start ___________（编辑图像）.
4. The ___________（保护开关）is at the front corner on the bottom of the scanner.
5. The scanner will take a few seconds to ___________（完成自检）.
6. The scanner will quickly perform a low-resolution scan of the original and the ___________（预览窗口）will show the preview image.

Unit 7 Computer Internet

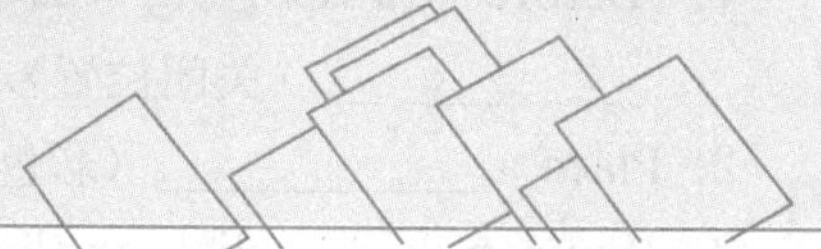

Learning Objectives

In this unit you should get familiar with:

- The functions of network marketing strategy;
- How the Internet works in market;
- The art of sales promotion;
- Understanding the form of the business letter.

Speaking

Dialogue 1

A: Hello, this is Pioneer Trading Company, Overseas Sales Department. May I help you?

B: Hello, I'd like to speak to Mrs. Jerry, please.

A: I'm sorry, I'm afraid she is on a business trip now, and she will come back next Monday. Can I take a message for you?

B: Yes, please tell her that I have got some problems on the merchandises, and I want to talk with her myself. Please ring back to me as soon as possible. My name is John Smith.

A: Could I take your contact?

B: Please call me this number: 22842772.

A: 22842772, all right, then I'll make sure she gets your message on her arriving.

B: Thanks for your trouble.

A: It's my pleasure.

Dialogue 2

A: Hello?

B: Hello. Is this 4474716? I'd like to speak to Mr. Wang, please?

A: I'm sorry. Mr. Wang is out right now.

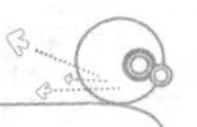

B: May I know when he'll be back?

A: I don't know, but he will certainly be back for lunch. This is his wife speaking. Can I take a message for him?

B: Thank you. Please tell him to be at the airport one hour before tomorrow afternoon.

A: Very good. I'll let him know as soon as he comes back. But, may I have your name, please?

B: This is Lin Ming. Thank you. Bye.

A: Good-bye.

Dialogue 3

A: Good morning.

B: Good morning. This is Li Gang. I'm calling from New York in America.

A: How can I help you?

B: I'm trying to get hold of Mr. Chen. Is he available?

A: I'll just find out for you, sir... Hello, are you still there? Unfortunately, Mr. Chen is not available at the moment. Would you like me to put you through to Mr. Li?

B: Yes, please. That's very kind of you.

A: You're welcome.

Dialogue 4

A: Hello!

B: Hello! Good morning. Who is that speaking?

A: It's Jane. I'm so sorry that I made such an early phone call.

B: It's nothing. Who do you wish to talk to?

A: Is Sue James in?

B: Sue, Jane wants you on the phone.

A: Hello! Is Sue there?

C: Yes, speaking.

A: Oh, sorry, I'm afraid I won't attend the meeting this morning. Last night I had a sore throat.

C: Do you have a temperature? Have you taken it?

A: No, haven't yet.

C: Don't worry about the meeting. You'd better go to see a doctor. I wish you will soon be well.

A: Thank you, Sue. Bye.

C: Bye.

Text Reading

Text A

Modern Enterprise Network Marketing Strategy

With the development of the Internet, how to use the cross-regional platform to break down barriers of resources, optimize the resource combination and improve the brand value, has become a hot talk and concerned research of many enterprises.

Product Strategy

Enterprises can determine the most suitable products for online sales by analyzing the characteristic of online consumers in general. And clear that the cost of the enterprise product sales on the Internet is far less than the cost of sales channels such as computer software and some other products. The online sales channel is more convenient than others, and naturally, the cost is much lower too, and this virtually reduced the enterprise cost, which improved the competitiveness of the enterprise products on the market.

Enterprises should take advantage of the opportunity that can communicate with customers directly, to provide customized products and services to customers. [1] At the same time, enterprises should understand consumer evaluation of enterprise products in a timely manner, so that enterprises can improve and accelerate new product research and development. Moreover enterprises can reduce the innovation risk and the development costs while developing the network market.

Price Strategy

Price is one of the most complex and difficult problems in the network marketing, because it is the most sensitive topic for enterprises, consumers and middlemen. Online marketing enables individual consumers can get more than one kind of product price and even all manufacturers' price, and then to make buying decisions, all of this decided that online sales price elasticity is bigger. Therefore, in order to make the most reasonable price, enterprises should give full consideration to check each link price formation in making online sales price.

Because online price will be affected by the impact of competition in the same industry at any time, enterprises can develop an automatic pricing system according to seasonal change, market supply and demand situation, competitive products, price changes, promotions and other factors, and adjust the actual price on the basis of calculating maximum profit. [2] At the same time, enterprises can also carry out market research, to obtain the relevant information to adjust the price.

Promotion Strategy

Online advertising is the common way of promotion. Internet advertising is different from other traditional advertising that widespread broadcast ("push"), but by consumer itself to choose ("pull"). The powerful features of the network contain almost all of the advantages of media advertising. Enterprises should give full play to the network multimedia acoustic-optic function and the three dimensional animation features in advertising planning, to induce consumers make purchase decisions as much as possible, and reach the goal that developing potential market. [3]

Using the chat function of network to carry out the consumer's fellowship activities or the online product exhibitions and promotional activities. This is a kind of promotion methods which mobilize consumer's emotional factors and promote emotional consumption. The online bookstore AMAZON is a typical success story in this aspect. It opened chat areas to attract readers, which make its annual sales reach 34% of increments, of which 44% is repeat business. As early as 1996 its sales broke through 17 million dollars, which displayed the charm of online promotion.

Make online promotion alliance with non competitive manufacturers through connecting the online database and the Internet, to increase the chances of contact with potential customers. On the one hand, this will make the enterprise products to be not impacted on, and on the other hand will widen the consumption level of the product. Integrate the network culture and product advertising, and attract consumers with the aid of the characteristics of network culture.

Channel Strategy

Combine with the related industry company, and build sale spots to sell series of products on the Internet together. Using this kind of means can increase the consumer Internet willingness and consumption motivation, and can also provide consumers with convenience, and increase the channel attraction. [4]

To set up a virtual store on enterprise website, and form a good online shopping environment through 3D multimedia design, and make a lot sorts of novel, personalized forms according to certain periods, season, promotional events, and consumer types to change store layout to attract more consumers into virtual stores.

There still have a lot of problems to solve in online trading, such as the security of online payment, but as the rapid development of Internet technology, online trading will becoming more and more perfect.

Notes:

[1] Enterprises should take advantage of the opportunity that can communicate with

customers directly, to provide customized products and services to customers.

企业应利用网络上与顾客直接交流的机会为顾客提供定制化产品和服务。

[2] Because online price will be affected by the impact of competition in the same industry at any time, enterprises can develop an automatic pricing system according to seasonal change, market supply and demand situation, competitive products, price changes, promotions and other factors, and adjust the actual price on the basis of calculating maximum profit.

因为网上价格随时会受到同行业竞争的冲击，所以企业可以开发一个自动调价系统，根据季节变动，市场供需情况，竞争产品，价格变动，促销活动等因素，在计算最大盈利基础上对实际价格进行调整。

[3] Enterprises should give full play to the network multimedia acoustic-optic function and the three dimensional animation features in advertising planning, to induce consumers make purchase decisions as much as possible, and reach the goal that developing potential market.

企业在做广告策划时，应充分发挥网络的多媒体声光功能、三维动画等特性，诱导消费者尽可能做出购买决策，并实现开发潜在市场的目标。

[4] Using this kind of means can increase the consumer Internet willingness and consumption motivation, and can also provide consumers with convenience, and increase the channel attraction.

采用这种方式可增加消费者的上网意愿和消费动机，同时也为消费者提供了便利，增加了渠道吸引力。

Vocabulary

cross-regional [krɒs, ˈriːdʒənl]	*adj*.	跨地区的
optimize [ˈɒptɪmaɪz]	*v*.	使最优化
virtually [ˈvɜːtʃuəli]	*adv*.	事实上
elasticity [ˌɪlæˈstɪsəti]	*n*.	弹性，灵活性
formation [fɔːˈmeɪʃən]	*n*.	形成，构造
acoustic-optic [əˈkuːstɪk, ˈɒptɪk]	*adj*.	声光的
dimensional [dɪˈmenʃənəl]	*adj*.	空间的
animation [ˌænɪˈmeɪʃən]	*n*.	卡通片绘制
increment [ˈɪŋkrəmənt]	*n*.	增量，增加
integrate [ˈɪntɪgreɪt]	*v*.	使完整

Exercises

Ⅰ. **Answer the following questions briefly according to the text.**

1. What are the network marketing strategies of the modern enterprises?

2. What has becoming a hot talk of many enterprises?

3. What are the advantages of network in promotion strategies?

4. How to increase the consumer Internet willingness and consumption motivation?

Ⅱ. Match the words on the left with their meanings on the right.

1. elasticity	A. modify to achieve maximum efficiency in storage capacity or time or cost
2. consumption	B. consider in detail and subject to an analysis in order to discover essential features or meaning
3. integrate	C. a prominent aspect of something
4. optimize	D. the tendency of a body to return to its original shape after it has been stretched or compressed
5. increment	E. of or relating to dimensions; having dimension—the quality or character or stature proper to a person
6. dimensional	F. a process of becoming larger or longer or more numerous or more important; the amount by which something increases
7. characteristic	G. make into a whole or make part of a whole; open (a place) to members of all races and ethnic groups
8. analyze	H. the process of taking food into the body through the mouth (as by eating; (economics) the utilization of economic goods to satisfy needs or in manufacturing

Ⅲ. Complete the sentences with the words from Ⅱ. Change the form if necessary.

1. "Although we live bathed in a sea of background radiation, people treat any ___________ as a dire risk," he said by e-mail.

2. The past, present and future only appear to those who exist within three ___________ reality.

3. Then the result set can be read and ___________ without holding any locks.

4. What are your goals in each of these spheres, and what can you do to ___________ both?

5. ___________ is the ability to scale up and down the infrastructure based on need.

6. Parents transmit some of their ___________ to their children.

7. The average annual electricity ___________ for a US residential utility customer is about 11, 000 kilowatt-hours.

8. You should consider them as an ___________ system in performance tuning.

Text B

To Be the Top Salesperson

Asking the wrong questions in cold calls can immediately ruin your chances of developing new businesses, while the right ones will open up opportunities, according to UK-based recruitment trainer, Roy Ripper.

Ripper, the director of recruitment, says that to get the most out of your prospecting calls, it's crucial to ask open-ended questions, for example:

- Can you tell me a little bit more about the nature of your work at Company X?
- How have you recruited in the past?
- Why would someone want to work for your company?

***Don't* say:**

- Are you recruiting at the moment?
- Do you have any other recruiters on your books?
- Would you say there's good potential for promotion at your company?
- Are there training courses available?
- Are you going to be recruiting again in the near future?

Reverse-Market Your Hottest Candidates

In the Recruitment Juice training DVDs, Ripper also provides a script for reverse-marketing candidates.

"What you *don't* want to do," he says, is call up and say, "Hi, it's your lucky day. I've got a candidate here who has four years' sales experience and is looking for a job."

The best way of introducing a candidate marketing call, according to Ripper, is to ensure that you sell the individual and what that person will do for that particular client.

He suggests saying something like: "My candidate has consistently been the top salesperson of the year in the last four years at his company; he's been promoted twice in that period and he's the only salesperson to be invited onto the company's fast track to management."

"The references for this guy are among the best I've ever seen, and he's expressed an interest in working for companies in your sector. So I'm approaching you and three of your competitors about him this afternoon. He's already got two interviews pencilled in for next week, so... when can I pencil you in for an interview?"

It's important to adapt any script to suit your personal style and your candidate's circumstances, because business development calls that "sound scripted" are among the top five hates of HR decision makers, Ripper says.

Other big hates, according to Recruitment Juice's research, are:

- Over-familiarity;

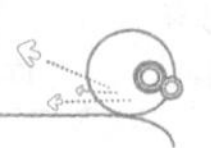

- Calling too often;
- An aggressive approach;
- Calling without a reason.

Business Development Tips

According to Ripper:

- Every call you ever make should be making an advance up the steps in your sales strategy;
- Find your optimum time of day, relevant to your marketplace, and aim to blast through your new business calls then;
- Prioritise your calls and make warm calls first, to build your confidence;
- Make the very first call you make every day a warm call;
- Spend at least 30 – 40 mins at the end of each day planning the next.

Vocabulary

cold call		向潜在的主顾打的电话
recruitment [rɪ'kruːtmənt]	*n.*	人员招聘
crucial ['kruʃəl]	*adj.*	重要的；决定性的
open-end ['əupən'end]	*adj.*	开放的
prioritise [praɪ'ɒrətaɪz]	*v.*	给予……优先权

Exercises

Ⅰ. Read the passage above and decide whether the following statements are true (T) or false (F).

( ) 1. According to UK-based recruitment trainer, Roy Ripper, asking the wrong questions in cold calls will immediately ruin your chances of developing new businesses.

( ) 2. You can ask some closed questions to get more information in prospecting calls.

( ) 3. According to Ripper, the best way of introducing a prospecting call is to ensure that you sell the individual and what that person will do for that particular client.

( ) 4. If you are going after the clients, you should call them consistently.

Ⅱ. Complete the following sentences by translating the Chinese in the brackets.

1. She said that she has good ________ at her company. (晋升潜力)
2. ________ are available in this company. (培训课程)
3. She's the only salesperson to be invited onto the ________ during these years. (公司管理层)

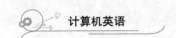

4. Your ________ are the best among these candidates.（推荐信）

5. Business development calls that "________" are among the top five hates of HR decision makers.（照本宣科）

Ⅲ. **Translate the following paragraph into Chinese.**

He suggests saying something like："My candidate has consistently been the top salesperson of the year in the last four years at his company；he's been promoted twice in that period and he's the only salesperson to be invited onto the company's fast track to management.""The references for this guy are among the best I've ever seen，and he's expressed an interest in working for companies in your sector. So I'm approaching you and three of your competitors about him this afternoon. He's already got two interviews pencilled in for next week，so... when can I pencil you in for an interview?"

Further Reading

HP's Marketing Strategy

The Products Strategy

Hewlett-Packard(HP) has a successful line of printers，scanners，digital cameras，calculators，PDAs，servers，workstation computers，and computers for home and small business use. Many of the computers came from the 2002 merger with Compaq. Hewlett-Packard now promote itself as supplying not justing hardware and software，but also a full range of service to design，implement，and support IT infrastructure.

Service Strategy

The goal of Hewlett-Packard service is to protect the customers' investment with award-winning service and support. Hewlett-Packard creates new possibilities for technology to have a meaningful impact on people，businesses，governments and society. There are diversity of services of different areas—technology services including support services and consulting services，enterprise services including application service，business process outsourcing，IT infrastructure outsourcing，cloud services and application transformation solution，additional services including support and drivers and education and training.

The Target Market and Customers

Hewlett-Packard，the Palo Alto，California-based information technology company，has been geared to a diversification strategy in recent years. The company wants to get more software-based products into its lineup. Such diversification helps companies guard against too much exposure to a specific product or industry. Not putting all their eggs in one basket means that if there is a downturn in one area，there is less of a chance that the

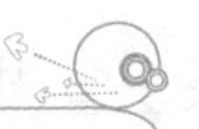

company will be impacted. Now Hewlett-Packard has a diversification of performance in products and services.

Hewlett-Packard has customers in every arena of the business world. To get the most out of its customers, the company began using the consumer as a guide to its small business and corporate initiatives. Their customers have shown a marked enthusiasm for this approach, which in turn has given Hewlett-Packard more qualified leads and business partners. This strategy allows the company to be led by the customer.

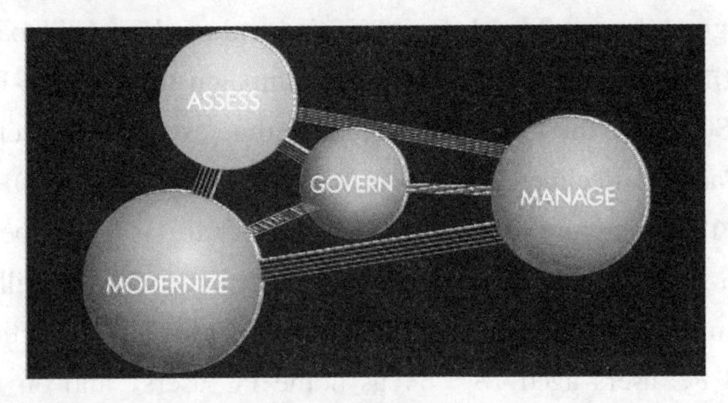

The Adaptations Made for the Chinese Market

Hewlett-Packard Co. is turning to China for engineers to develop global storage and networking products, opening a new research center in Beijing this year as it seeks to boost sales in emerging markets.

In China, the company's main focus would be in cloud computing and social media. The company didn't provide much detail about the size of the investment, but the former Chief Executive Officer Leo Apotheker, is bullish on the region. The CEO was also of the opinion that China's vibrant economy, with high growth rate in social and mobile connectivity, and strong commitment to innovation, which presents tremendous opportunities for Hewlett-Packard.

Future Plans & Goal

Hewlett-Packard's reliance on a Customer Relationship Management program is no accident. By creating goals that can be measured, Hewlett-Packard has managed to cut spending by 10 percent. These goals allow the company to funnel the marketing dollars to the strategies that show tangible results, creating a more efficient and effective marketing strategy.

Hewlett-Packard will continue to move aggressively to drive growth, expand margins, and deliver unparalleled value to the consumers and stockholders.

The company plan to increase investments in innovation and will grow by investing in portfolio enhancements and acquisitions, by covering more of the market with our sales force and channel partners, and by expanding that courage more aggressively into high-growth emerging economies.

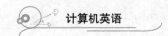

The company will continue moving up the technology stack into higher margin categories.

Advertising Strategies

Hewlett-Packard's strategy was to listen to customers, to understand their needs, to deliver solutions, to translate them into customer success and to earn customer loyalty. Hewlett-Packard's log is 'Hewlett-Packard invent'. In 2006 Hewlett-Packard implemented new global advertising strategy under the message: "The Computer Is Personal Again". This advertising campaign embraces TV, Internet, print and outdoor (billboards). Neal Leavitt claims that this is the first ever global marketing campaign for Hewlett-Packard's "Personal Systems Group (PSG) and the company is hoping that expending several hundred million dollars (Hewlett-Packard declined to divulge total ad campaign costs) will not only help bolster PC sales (the company's worldwide market share is 15.7 percent according to research firm IDC; market leader Dell, Inc. has 18.2 percent), but will also help reinforce its approach to empowering users and enterprises" (iMedia Connection 2007). Target audience embraced PC users aged 18 - 34 as home PC users, and on the business front, small and middle sized companies.

Major Competitors of the Industry

Dell, IBM and Accenture company are three major competitors of Hewlett-Packard in IT industry. In China, Hewlett-Packard is top 5 computer companies. They compete Hewlett-Packard in computer equipment, consulting, technology solution, software and many other areas of products and services.

Writing

Business Letter

商业信函是用正式语言书写的信函，通常是一个商业机构写给另一个商业机构的书面信函，或者是这些商业机构与客户或其他外部机构之间的信函。它包含公司信息及商务事件的详情。

The Structure of a Business Letter

1. letterhead 信头
2. reference and date 编号和日期
3. inside name and address 封内名称和地址
4. attention line 经办人
5. salutation 称呼
6. subject line 事由
7. body 正文
8. complimentary close 结尾敬辞

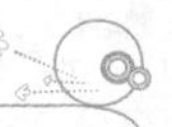

9. signature 签名

10. enclosure 附件

11. carbon copy notation 抄送

12. postscript 附言

Sample

> **H. Woods & Co. Ltd.**
> Nesson House，Newell Street
> Birmingham B15 3EL
> Tel：(44-121) 4560000　Fax：(44-121) 4560001
>
> 1 March，2001
> The Manager
> Shanghai Textile Trading Co. Ltd.
> 72 Zhongshan Rd.
> Shanghai 200001
> China
> Dear Sir or Madam，
> We are interested in tweed lengths suitable for skirt-making and would like to have details of your prices and terms.
> It would be helpful if you could supply samples.
> Yours faithfully，
> Larry Crane
> China Trade Manager

Task：Please write a business letter according to the following information.

这是一家南京电器用品进口商向波士顿电子公司提出进口业务要求的信函。写信人首先说明他们从因特网上得知对方的业务范围碰巧和自己公司相同，因此写信想要建立合作关系。接着简单介绍了其公司的业务状况并给对方提供了资信调查途径。最后提及对对方无绳电话的兴趣及对回复的期待。

Unit 8 Electronic Business

Learning Objectives

In this unit you should get familiar with:

- The main idea of electronic business;
- The transactional mode of the eBay business;
- The ways to make money with network trade;
- The writing skills of invitation.

Speaking

Dialogue 1

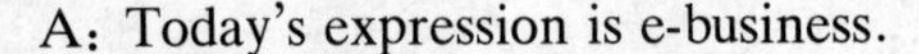

A: Today's expression is e-business.

B: E-business?

A: Well, e-business is electronic business. It means any business done using the Internet.

B: That's right. You could say e-business has grown very much in China recently.

A: That's right.

Dialogue 2

A: Why did you close your office?

B: Well, it's all e-business these days, so I can just work from home on my computer.

A: Do you think we should start an e-business service?

B: Why not? It could bring in sales in 24-7. I'm sure our customers would love it.

Dialogue 3

A: So Helen, what type of companies use e-business?

B: Well, anything to do with computers can be e-business. A lot of travel companies use e-business, too.

A: Yes, that's true, I have used online booking systems.

B: Yes. It is very easy to buy plane tickets or reserve a hotel room on the Internet.

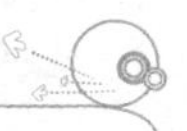

A: But you have to be careful with your credit card details.

Dialogue 4

A: I thought computers were supposed to make your life more convenient. But it seems like I spend half my time waiting for it to do something.

B: Why is it so slow? How much RAM do you have?

A: I think I have 8 Megs.

B: That's your problem. If you want Windows 95 to run smoothly, you need at least 16 Megs.

Dialogue 5

A: I have 16 Megs now. And most things work better, but my Internet connection is still really slow. Is there anything that I can do?

B: Let's see. Your modem is only a 14.4. If you want your Internet connection to speed up, you'll have to get a faster modem.

Dialogue 6

A: I can't seem to get my mouse to work properly. First it moves and then, all of a sudden, it jerks all the way across the screen. What can I do?

B: Here, let me see. Open the mouse and takes out the ball. Do you see these contacts here?

A: Yes. They look pretty dirty.

B: That's right. Let me clean those for you, and it'll be back to normal in no time.

Text Reading

Text A

What Is Electronic Business (EB)

Electronic Business (EB) means companies make use of Internet and information technologies, aiming at raising operation efficiency, to carry out every phase of business procedures automatically, electronically and with paper documents eliminating. These phases include: market researching, negotiating, delivering, producing, money transferring and after-sale services providing and so on.[1]

EB is not a kind of pure technique itself, but an application process of Internet and information technologies in doing business. In such process, the tools used for information exchanging, transmitting, processing and storing based on paper media are replaced by the tools based on electronic media.[2]

In traditional business surroundings, a company should spend quite a lot of time and effort to investigate and decide what and how to produce (or sell), then, advertise in mediums (such as TV, radio, bulletin board on street and so on), contact the buyer through phone, fax, mail or in face to face after looking at the goods, the buyer bargains, places the order or signs the agreement with the seller. The seller prepares the goods and finally sends the goods to the buyer. The buyer pays for the deal in cash or by check. In the above procedure all information is exchanged by phone, mail, fax or telegraph, and interest decades—in EDI (Electronic Data Interchange). EDI is a computer network system dealing with some special data exchange such as invoices, contracts, credit letters and so on between certain partners. To build an EDI system costs a lot of money and it is rather complicated to use it. So, as you can see, in traditional business surroundings, people spend more time and efforts to do business. Sometimes mistakes and delays will happen.

In EB surroundings, people use Internet or WWW to do business. The contact between the buyer and the seller has far more chances to know each other, not only one to one, but many to many. No geographical constraints, no time limited, all of them can find satisfied partners. The buyer uses browser to seek out the product list on the home page of the website or online shops based on the wanted domain names, and see the packing, outlook, style, color and price of products, sometimes, then can bargain online. If he is satisfied with the goods, he can place an order by click the key board or by e-mail. The seller immediately receives the message. If he accepts the order, he can use Internet or WWW to arrange goods production, distribution as well as payment (called electronic funds transfer). All data transmitted just in time and business operation becomes more efficient.

Vocabulary

electronic [ɪlek'trɒnɪk]	*adj.*	电子的，电子操纵的
efficiency [ɪ'fɪʃənsi]	*n.*	效率，效能
procedure [prə'siːdʒə]	*n.*	程序，过程
eliminate [ɪ'lɪmɪneɪt]	*v.*	消除，排除，淘汰
transmit [træns'mɪt]	*v.*	发射，播送，广播
investigate [ɪn'vestɪgəɪt]	*v.*	调查，审查
medium ['mɪdɪəm]	*n.*	媒介，手段，方法
invoice ['ɪnvɔɪs]	*n.*	发票
complicated ['kɒmplɪkeɪtɪd]	*adj.*	结构复杂的
geographical [dʒɪə'græfɪkl]	*adj.*	地理学的，地理的
constrain [kən'streɪn]	*v.*	强迫，强使
browser ['braʊzə]	*n.*	浏览程序，浏览器
distribution ['dɪstrə'bjʊʃən]	*n.*	分发，分配，（商品）运销

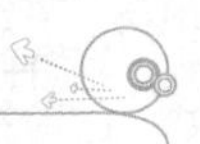

Phrases and Expressions

aim at	（以……）瞄准，以……为目标；计划；打算
carry out	执行，贯彻，完成，实现
seek out	找出
be satisfied with	对……感到满意

Notes:

[1] Electronic Business（EB）means companies make use of Internet and information technologies，aiming at raising operation efficiency，to carry out every phase of business procedures automatically， electronically and with paper documents eliminating. These phases include：market researching，negotiating，delivering，producing，money transferring and aftersale services providing and so on.

电子商务是指企业利用因特网与信息技术，以提高经营效率为目标，在企业经营全过程的各个环节实现自动化、电子化、无纸化，这些环节包括市场研究、谈判、配送、生产、资金支付和售后服务等。

[2] EB is not a kind of pure technique itself，but an application process of Internet and information technologies in doing business. In such process，the tools used for information exchanging，transmitting，processing and storing based on paper media are replaced by the tools based on electronic media.

电子商务本身不是一种纯技术，而是因特网与信息技术在商务中的运用过程，以纸介质为基础的用于信息交流、传递、处理及存储的工具被以电子媒体为基础的工具所取代。

Exercises

Ⅰ. **Read the passage above and decide whether the following statements are true（T）or false（F）.**

（ ）1. EB is a kind of pure Internet and information technologies in doing business.

（ ）2. In traditional business surroundings，people spend more time and efforts to do business.

（ ）3. In EB surroundings，the buyer and the seller have little chance to know each other.

（ ）4. The buyer can't see the packing，outlook，style，color of goods online.

（ ）5. If the buyer is satisfied with the goods，he can place an order by click the key board or by e-mail.

Ⅱ. **Fill in each blank with the appropriate form of the words given below.**

efficiency, constrain, eliminate, complicated, medium, invoice be satisfied with, aim at, seek out, carry out

1. Working at peak __________.
2. Broil a (an) __________ steak.
3. Having __________ nutritional requirements.
4. The factory must __________ increasing production.
5. One shouldn't __________ only a little success.
6. She tried to __________ herself from a cough in class.
7. I make out a (an) __________.
8. __________ big orders from department stores.

Ⅲ. **Match the following terms with their Chinese meaning.**

A	B
1. operation efficiency	A. 售后服务
2. market researching	B. 纸介质
3. money transferring	C. 资金转移
4. after-sale services	D. 市场调研
5. paper media	E. 经营效率
6. in cash	F. 时间限制
7. by check	G. 地域限制
8. traditional business	H. 电子资金转移
9. geographical constrain	I. 传统经营
10. time limited	J. 称心如意的对家
11. satisfied partner	K. 现金支付
12. electronicfunds transfer	L. 支票支付

Ⅳ. **Translating.**

A. Translate the following passage into Chinese.

In EB surroundings, people use Internet or WWW to do business. The contact between the buyer and the seller has far more chances to know each other, not only one to one, but many to many. No geographical constrains, no time limited, all of them can find satisfied partners.

B. Translate the following sentences into English.

1. 管理人员能从浏览器上配置设备。(browser)
2. 我希望您能对第一批货感到满意。(be satisfied with)
3. 政府能消除贫穷吗?(eliminate)

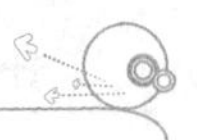

4. 警察用马制止群众接近会场。(constrain)
5. 征求读者的批评和意见。(seek out)

Text B

The EBay Business

In September 1995, Pierre Omidyar created the online auction conglomerate known as eBay. His vision was to create a virtual marketplace for the sale of goods and services for and by individuals.

To get the business off the ground, Omidyar and cofounder Jeff Skoll enlisted the help of Margaret Whitman, a famous Harvard Business graduate. Whitman, who formed her management team with senior-level executives from industry leaders like Pepsico and Disney, helped create a strong mission for eBay, which was to be in the business of connecting people—not selling products to them. Rather, they've created a person-to-person marketplace on the Internet, where sellers list items for sale and interested buyers bid on these items.

When eBay was first launched, the site immediately became a popular channel for auctioning collectibles, but it soon developed into other markets such as automobiles, business and industrial equipment, and consumer electronics, where the average sales price (ASP) is higher. Increasing the ASP was an important component of increasing sales for eBay, as eBay's transaction fees are based on a percentage of sales. The higher the ASP, the more money eBay makes on each of the million transactions it facilitates.

EBay has pioneered and internationalized automated online person-to-person auctioning. Previously, such commerce was conducted through garage sales, collectibles shows, flea markets, and classified advertisements. An online marketplace facilitates easy skimming for buyers and enables sellers to list an item for sale within minutes of registering.

Browsing and bidding on auction is free, but sellers are charged transaction fees for the right to sell their goods on eBay. There are two kinds of transaction fees: When an item is listed on eBay, a nonrefundable insertion fee is charged based on the seller's opening bid on the item. Once the auction is completed, a final value fee is charged. This fee generally ranges from 1.25 percent to 5 percent of the final sale price.

EBay also raises its listing fees with improved auction features, including attractive or bold listings, featured status, and other ways for sellers to increase the visibility of their items. Once the auction is finished, eBay notifies the buyer and seller via e-mail. Completing the transaction is then up to the seller and the buyer, and eBay collects its final value fee independent of payment and shipment.

Vocabulary

auction [ˈɒkʃən]	*n*./*v*.	拍卖
conglomerate [kənˈglɒmərət]	*n*.	联合企业，企业集团
collectible [kəˈlektəbl]	*adj*.	可收集的
browse [braʊz]	*v*.	浏览
nonrefundable [ˌnɒnrɪˈfʌndəbl]	*adj*.	不可退还的；不归还的

Phrases and Expressions

get… off the ground	使……开始起步
person-to-person marketplace	人对人的贸易市场
collectible show	收藏品展览
classified advertisement	分类广告
insertion fee	注册费；登记费
final value fee	最终交易费

Exercises

Ⅰ. Read the passage above and decide whether following statements are true(T) or false(F).

( ) 1. In September 1995, Pierre Omidyar founded the online auction conglomerate known as eBay, which is a virtual marketplace.

( ) 2. The aim of creating eBay is to sell products to people who need.

( ) 3. At the outset, eBay cannot attract peoples' eyeballs.

( ) 4. It is free of charge for sellers to browse and bid on auction on eBay, but when they sell their goods on eBay, the transaction fees are charged.

( ) 5. Once the auction is completed, a nonrefundable insertion fee is charged, generally ranging from 1.25% to 5% of the final sale price.

( ) 6. EBay has pioneered and internationalized automated online person-to-person auctioning.

Ⅱ. Complete the following sentences by translating the Chinese in the brackets.

1. ____________（为了使 eBay 贸易起步），Omidyar and cofounder Jeff Skoll enlisted the help of Margaret Whitman, a Harvard Business graduate.

2. When eBay was first launched, the site immediately ____________（成为一种时尚的商品拍卖渠道）.

3. In comparison with eBay, the previous commerce was conducted through ____________（库房销售）, ____________（收集物展览）, ____________（跳蚤市场）and ____________（分类广告）.

4. The two kinds of transaction fees for the sellers to sell their goods on eBay are

 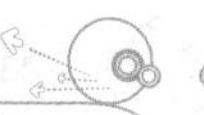 

charged，one is _________ （不退还的注册费），and the other is _________ （最终交易费）.

5. Once the auction is finished，eBay ___________ （通过电子邮件通知买卖双方）.

Further Reading

Seven Ways to Make Money with Electronic Trade

As e-commerce develops further，the Web has become an intensely commercial medium，which offers plenty of ways to make money. Here are some suggestive points for you to try：

Sell advertising and sponsorships for your site. This is the classic Web business model，selling “eyeballs” to advertisers who want to reach the audience your site attracts and charging for ads on their sites in terms of time.

Sell products and services from your site. Nowadays commerce is king on the web as everyone scrambles（抢夺，争夺）to enable e-commerce on their websites though doing it right is harder than it looks.

Point people to other sites. This may be simplest and easiest way to make money with your site. Simply point visitors to your site to a related book or other products on Amazon. com and you collect a commission（佣金）on anything the user buys.

Collect and sell information. Creating a database of your users isn’t just a good idea—it’s good business. Demographic（统计的）information about your users is a potential gold mine，and specific information about a user’s preferences and interests is even more valuable.

Charge a free access to your site—or at least part of your site. Even though the web offers a cornucopia（大量）of free content，if your content is unique and valuable，you can make money charging for it online. Some newspapers，for example，offer free online access to their daily editions but charge to view their archives（档案，案卷）.

Do e-mail marketing. E-mail is an even more lucrative（赚钱的）revenue（收入）source than the Web. You can sell ads on e-mail newsletters（简报，简讯），and use targeted e-mail to alert customers about special deals or new products. Just lay off the spam，please—make sure that a customer wants your e-mail before you send it.

Rent software online. Some businesses now supply software on a “pay per use” basis. Instead of licensing（授权，允许）software on a permanent basis，users can run it over the net，whenever they need it，from your servers. Businesses that rent software don’t have to pay the fulfillment costs associated with standard software sales，and they also enjoy the benefits of an ongoing relationship with their customers.

Writing

Invitation Letter&Reply

邀请信分为两种：一种属于个人信函，如邀请某人共进晚餐、参加宴会、观看电影、出席典礼等；另一种邀请信则属于事务信函，一般是邀请参加会议、学术活动等。

一、普通邀请信

第一种邀请信邀请的对象一般是朋友、熟人，所以内容格式上的要求都比较松，可以写得随便一些，只要表明邀请的意图，说明活动的内容、时间、地点等。但既然是邀请信，那么就一定要在信中表达非常希望对方能够参加或者出席的愿望。这种邀请信的篇幅可以非常短，下面以一封邀请看电影的短信为例。

Dear Jane,

We have four tickets for a famous film shown at Guangming Cinema, *The Longest Day*, Friday, the ninth. Will you join us? We'll be looking for you at eight sharp Friday night in front of the cinema, so don't disappoint us!

Warmest regards.

Alice

二、正式邀请信

第二种邀请信一般由会议或学术活动的组委会的某一个负责人来写，以组委会的名义发出，而且被邀请者通常也是属于比较有威望的人士。因此，这类邀请信的措辞要相对正式一些，语气要热情有礼。

这一类邀请信通常要包括以下内容：首先表明邀请对方参加的意图以及会议或学术活动的名称、时间、地点；然后要对被邀请者的威望和学术水平等表示推崇和赞赏，表明如果被邀请者能够接受邀请，会给会议或者活动带来很好的影响；接着要说明会议或活动的相关事宜，最好是能引起对方兴趣的事宜；当然不能忘了表达希望对方能够参加的诚意；最后，还要请收信人对发出的邀请做出反馈，如确认接受邀请。下面是一封邀请对方参加学术会议的信函。

Dear Professor Wang,

On behalf of the Ohio State University and the IEEE Computer Society, I would be very pleased to invite you to attend and chair a session of the forthcoming 2004 International Conference on Parallel Data Processing to be held in Bellaire, Michigan, from October 25 to October 28, 2004.

You are an internationally acclaimed scholar and educator. Your participation will be among the highlights of the Conference.

We sincerely hope that you could accept our invitation. As you know, this is the 10th anniversary of the Conference and we plan to make it a truly international meeting. We have accepted many papers from several foreign countries, including two from China.

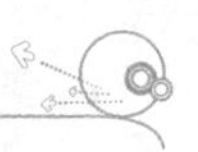

If you can come, please let us know as soon as possible, since we have to prepare the final program soon. We are looking forward to your acceptance.

Sincerely yours,
Peter White

三、邀请信的回复

无论是收到哪种邀请信都要回信，明确表示接受与否，而且无论接受与否都要对邀请者表示感谢。接受邀请的回信一般包含以下内容：首先，感谢对方的邀请，并重述邀请信中的主要内容；然后，愉快地表示接受邀请，并简要说明自己的打算；最后，表示期待赴会和与对方见面的心情。下面是一封接受邀请的信，与上面第一封邀请信相对应。

Dear Alice,

Thank you very much for your invitation. It will be great pleasure for me to join you on Friday for the wonderful film. I will arrive at the cinema before eight. I look forward to meeting you on Friday.

Thank you for thinking of me.

Yours,
Jane

谢绝邀请的回信一般包含以下几方面的内容：首先，还是要对对方的邀请表示感谢；然后，具体说明自己无法接受邀请的原因，并对无法出席表示遗憾；最后，表达自己的祝愿，即祝愿自己未能参加的会议或者活动能够顺利进行。下面是一封谢绝邀请的信，与上面的第二封邀请信相对应。

Dear Professor White,

Many thanks for your letter dated 15th August, inviting me to attend and chair a session of the forthcoming 2004 International Conference on Parallel Data Processing to be held in Bellaire, Michigan, from October 25 to 28, 2004.

Much to my regret, I shall not be able to honor the invitation because I have been suffering from a disease since this summer. I am firmly advised that it would be unwise to undertake any distant and long travel in the near future.

I feel very sad to miss the opportunity of meeting you and many others in the field of Computer Science. I wish the conference a complete success.

Faithfully yours,
Wang Xuan

Task: Write an invitation letter in the following situations.

给你的朋友 Jenny 写一封邀请信，邀请她于 12 月 25 日晚八点和你一起去北京音乐厅听圣诞音乐会。

Unit 9 Computer Security

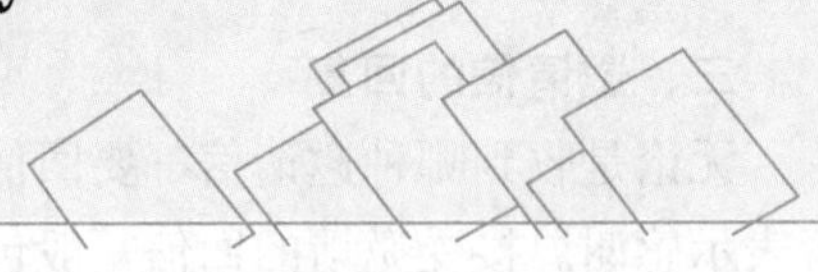

Learning Objectives

In this unit you should get familiar with:

- The importance of computer security;
- The security measures of electronic business;
- How to protect our own data on the Internet;
- Understanding and filling in agenda.

Speaking

Dialogue 1

A: What can I do for you?

B: The system crashed when I was surfing on the Internet.

A: Did you go to any illegal website?

B: No. But does that matter?

A: Yes. Your computer can be easily infected by virus if you do that.

B: I see. I'd better never try.

A: That's wise.

Dialogue 2

A: Do you know what's wrong with my PC?

B: One minute. Oh, yes, it was infected by a virus, and you had no anti-virus software.

A: Is anti-virus software necessary for a PC?

B: Of course. You'd better learn something about it.

A: I'm afraid yes. But what about the data I stored in the computer?

B: Don't worry, it should have been protected automatically. And I take an anti-virus software with me. Do you want me to install it now?

A: Yes, please. I'll really appreciate that.

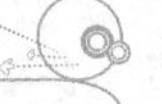

Dialogue 3

A: Do you have virus protection software on your PC?

B: No, I'm always installing new software, and getting rid of old programs. Virus protection software usually makes that more of a hassle.

A: But what will you do if you get a virus?

B: I guess I'll just reformat my hard drive and start over.

Dialogue 4

A: Do you use an anti-virus program to protect your computer?

B: Yes, I do. I also use an anti-spyware program and a firewall. There are so many nasties on the Internet and so many people who are trying to use the Internet to hurt other users. You have to be very careful and keep you computer clean.

A: How often do you run your anti-virus program?

B: I usually run it every few days. It finds a virus about half the time.

Dialogue 5

A: Which anti-virus program would you recommend?

B: Have you tried this one? It's very good and you can download it for free on the Internet. You can also download updates for free.

A: That sounds very good. Which website should I visit to download it?

B: Just a minute. Go to this one. I'll send you an e-mail with the links in it. It only takes a few minutes to download. Then you have to go through the set up procedures. But they are not complicated.

Dialogue 6

A: Help! I just found out that I have a virus on my computer!

B: Me, too. That's why I have two anti-virus softwares installed.

A: I didn't have anti-virus software installed, but I never got virus.

B: Really? How did you manage that?

A: I never connect my computer to the Internet. And I don't borrow other people's discs.

B: I see. So, what are you going to do now?

A: It seems I have to send it to the computer shop.

B: I know a friend of mine is really good with computers. Maybe I can ask him to help you out?

A: That's so kind of you.

Text Reading

Text A

The Security of EB

Something Worried in EB Application

The Internet brings people endless convenience and opportunities, and at the same time, it also brings risks to them. Since EB is conducted on Internet, and the Internet is such an open system, so when data must be electronically processed, transmitted and stored in some modes, some risks occurred. The risks associated with the Internet and EB applications may occur in many different ways. For example, data may be stolen, corrupted, misused altered or falsely generated. Computer system or website may be attacked and make render systems unable to operate properly, hardware or software is used illegally. And these may make the enterprise users suffered serious damage or ruin their business.

Why are so many unhappy things happened? There are two reasons. One reason is that, the Internet is a new generation, it is developing very fast. Its attractive perspective, multifunctional service and exponential growth rate of web sites opened on it attract more and more companies, and private users to join in just like to join a gold rush. But in many aspects, such as the information technologies, the protocols, the language, the standards, polices and laws, which are the premise of Internet operations, are far from integrity. [1]

The other reason is that people are so eager to join the Internet and EB applications that they haven't enough time to fully think over the details that may involve in. So they can't be ready at any time to deal the internet perpetrators a head-on blow, for they haven't enough knowledge and experiences about the risks they may encounter, even they don't know how to detect the attempts of these perpetrators. [2] So they haven't an adequate control over these perpetrations.

The heavy lessons of Internet perpetration draw people's attention to improving the situation of Internet and EB security, for they expect to conduct their electronic business in a secure circumstance. [3] Generally, the basic requirement for EB security can be conclude like this:

First of all, secure the privacy of electronic transaction. That means all the electronic data are protected from unauthorized disclosure. There are many kinds of transaction data, for instance, advertisements and product lists, customer's personal files and confidential data. So we should take different methods to secure different kinds of data. Another meaning of privacy is that the electronic transactions and their contents

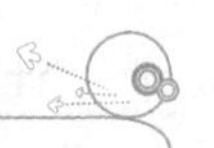

should and could not be known by any person, except the trading partners. Currently, the security techniques used for secure the privacy are encryption, firewall and passwords.

Second, secure the integrity of electronic transactions. This means to secure and verify the data elements and its contents captured in an electronic transaction base on the agreed elements, and maintain the integrity of the data elements in processing and storage procedures and do not be altered in any unauthorized fashion.

Third, the security assurance systems. When the user receives the message sent electronically by a sender, both the user and sender's identity need to be conformed in order to verify whether the sender or receiver is he /she claims to be.[4] Generally, their identities could be proved through checking their passwords, digital signatures or digital certificates issued by the authorized third party. Another thing need to assure is whether the transaction exists, which is called "non-repudiation".

In conclusion, the EB security requirement is that the content and privacy of electronic transaction must be protected from being intercepted, abused, altered, deleted or disturbed during the electronic data interchange processing, transmitting and storing.

Notes:

[1] But in many aspects, such as the information technologies, the protocols, the language, the standards, polices and laws, which are the premise of Internet operations, are far from integrity.

但在许多方面，如信息技术、网络协议、语言、网络标准、制度与法规等网络运作有关的方面远远没有完善。

[2] So they can't be ready at any time to deal the Internet perpetrators a head-on blow, for they haven't enough knowledge and experiences about the risks they may encounter, even they don't know how to detect the attempts of these perpetrators.

所以他们根本不可能随时做好准备以给网上犯罪者迎头痛击，因为他们没有足够的知识与经验应对他们可能面对的那些风险，不知如何发现那些犯罪者的企图。

[3] The heavy lessons of Internet perpetration draw people's attention to improving the situation of Internet and EB security, for they expect to conduct their electronic business in a secure circumstance.

网上犯罪的沉痛教训将人们的注意力转向改善网络及电子商务安全状况，因为他们期望在一个安全环境下处理他们的电子商务。

[4] When the user receives the message sent electronically by a sender, both the user and sender's identity need to be conformed in order to verify whether the sender or receiver is he /she claims to be.

当合作伙伴送来的电子信息被用户接收时，用户和发送者的身份都需要证明以确认接、收双方是否是他/她所自称的那个人。

Vocabulary

endless ['endlɪs]	*adj*.	无止境的；连续的
process ['prəʊses]	*v*.	处理；加工
transmit [trænz'mɪt]	*v*.	传播，传导
mode [məʊd]	*n*.	模式；方式
corrupt [kə'rʌpt]	*v*.	堕落，腐化；腐烂
misuse [mɪs'juːz]	*v*.	滥用；误用
illegally [ɪ'liːgəli]	*adv*.	非法地
enterprise ['entəpraɪz]	*n*.	企业；事业
perspective [pə'spektɪv]	*n*.	观点；远景
multifunctional [mʌltɪ'fʌŋkʃənəl]	*adj*.	多功能的
exponential [ˌekspə'nenʃ(ə)l]	*adj*.	指数的
protocol ['prəʊtəkɒl]	*n*.	协议；草案；礼仪
premise ['premɪs]	*n*.	前提
integrity [ɪn'tegrɪtɪ]	*n*.	完整；正直；诚实
perpetrator ['pɜːpətreɪtə(r)]	*n*.	犯罪者；作恶者
encounter [ɪn'kaʊntə]	*v*.	遭遇，邂逅；遇到
detect [dɪ'tekt]	*v*.	察觉；发现
adequate ['ædɪkwət]	*adj*.	充足的；适当的
perpetration [ˌpɜːpɪ'treɪʃən]	*n*.	犯罪；做坏事
circumstance ['sɜːkəmstəns]	*n*.	环境，情况
privacy ['prɪvəsɪ; 'praɪ-]	*n*.	隐私；秘密
transaction [træn'zækʃ(ə)n]	*n*.	交易；事务
unauthorized [ˌʌn'ɔːθəraɪzd]	*adj*.	非法的；未被授权的
disclosure [dɪs'kləʊʒə]	*n*.	披露；揭发
confidential [kɒnfɪ'denʃ(ə)l]	*adj*.	机密的；信任的
encryption [ɪn'krɪpʃən]	*n*.	加密；加密术
verify ['verɪfaɪ]	*v*.	核实；查证
assurance [ə'ʃʊər(ə)ns]	*n*.	保证，担保
signature ['sɪgnətʃə]	*n*.	署名；签名
intercept [ˌɪntə'sept]	*v*.	拦截；窃听

Phrases and Expressions

associate with	联合；与……联系在一起
head-on blow	突然而来的打击；当头棒喝

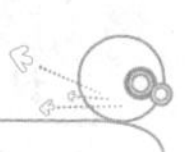

Exercises

Ⅰ. Answer the following questions briefly according to the text.

1. Why some risks occur when data must be electronically processed, transmitted and stored in some modes?
2. What risks people may meet when EB is conducted on Internet?
3. What area does the lessons of Internet perpetration draw people's attention to?
4. What does securing the privacy of electronic transaction mean?
5. What's the basic requirement for EB security?

Ⅱ. Fill in each blank with the appropriate form of the words given below.

associate with, illegal, generate, transmit, integrity, disclose, adequate

1. Cigarette smoking has been __________ lung cancer.
2. The state places great hopes on the younger __________.
3. It depends on the __________ of the satellite signals.
4. For criminal attacks, the motivation is to gain money __________.
5. The news' __________ shocked the public.
6. I'm sure her knowledge of English is __________ for the job.
7. We must safeguard our state territorial __________.

Ⅲ. Fill in the table below by giving the corresponding Chinese or English equivalents.

	爆炸式增长
head-on blow	
	电子商务应用
electronic transaction	
unauthorized disclosure	
	机密数据
	防火墙
security assurance system	
	数字签名
digital certificate	

Ⅳ. Translating.

A. Translate the following sentences into Chinese.

1. The Internet brings people endless convenience and opportunities, and at the same

time, it also brings risks to them.

2. Its attractive perspective, multifunctional service and exponential growth rate of web sites opened on it attract more and more companies, and private users to join in just like to join a gold rush.

3. The other reason is that people are so eager to join the Internet and EB applications that they haven't enough time to fully think over the details that may involve in.

B. Translate the following sentences into English.

1. 和积极的人们交往，保护自己不受所有形式的消极侵害。(associate with)
2. 正因为我想在将来成为一名教师，所以我满怀好奇来做这个研究。(be eager to)
3. 周六的事故对我们的计划是一个迎头打击。(head-on blow)
4. 针对这些问题，我们也可以采取一些行之有效的方法予以解决。(take method to)
5. 这个项目在带来利益的同时，也隐藏着巨大的风险。(bring risk)

Text B

Computer Network Attacks and Prevent

Network security is the security of computer information system in an important aspect. As the opening of Pandora's Box, the computer systems of the Internet, greatly expanded information resources sharing space at the same time, will be exposed in their own malicious more under attack. How to ensure the network storage security, and the transmission of information security is the so-called computer network security. Information security, including operating system security, database security, network security, virus protection, access control, encryption and identification these seven areas.

Here are some network security solutions.

First, the deployment of intrusion detection system.

Ability of intrusion detection is an important factor of an effective defense system. Powerful integrated intrusion detection system can make up for the inadequacy of relatively static defense of firewall, do real-time detection for the various acts of the external network and the campus network, and from the real-time detection, it will detect all possible attempts to attack and take corresponding measures.

Second, vulnerability scanning system.

Using the most advanced vulnerability scanning system scanning workstations, servers, switches and other safety checks regularly, and in accordance with the results to provide the system administrator with detailed and reliable security analysis, in order to enhance the overall level of network security have an important basis.

Third, the network version of the deployment of anti-virus products.

In the network anti-virus program, we eventually have to reach the goal to put an end to the LAN in the whole virus infection, dissemination and attack. In order to achieve this, we should take the corresponding anti-virus tools in the entire network which is at risk of infection and the spread of the virus to take place. At the same time, in order to implement and manage the entire network of anti-virus system effectively and quickly, it should be able to achieve remote install, smart upgrade, remote alarm, centralized management, distribution killing and other functions.

Forth, the networkhosts operating system security and physical security measures.

Network firewall as the first line of defense can not protect the internal network fully. It must be combined with other measures to improve the level of the safety system. In accordance with the level from low to high, namely, the physical security of the host system, the core operating system security, system security, application services security and file system security. At the same time, host security checks, bug fixes, as well as a backup safety system as a supplementary safety measures, these constitute the entire network system. The second line of defense is the main part of a breakthrough to prevent the firewall as well as attacks from within. System backup is the last line of defense network system, used to attack after the System Restore. The firewall and host security measures is the overall system security by auditing, intrusion detection and response processor constitute the overall safety inspection and response measures.

From the network system firewall, network host or even directly from the network link layer on the extraction of network status information, and input to the intrusion detection subsystem. Intrusion detection system in accordance with certain rules to determine whether there is any invasion of the incident. If the invasion occurred, it will take emergency treatment measures, and generate a warning message. Moreover, the system's security audit, also can be used as the future consequences of aggressive behavior and to deal with security policy on the system to improve sources of information.

On the whole, network security is a comprehensive issue. It involves technology, management, use and many other aspects, including both its own information system security issues, and physical and logical technical measures. As a kind of technology, it can only solve the problem on the one hand, rather than a panacea.

Vocabulary

Pandora [pæn'dɔːrə]	*n*.	潘多拉
expand [ɪk'spænd; ek-]	*v*.	扩张；使膨胀
expose [ɪk'spəuz; ek-]	*v*.	揭露，揭发
malicious [mə'lɪʃəs]	*adj*.	恶意的；恶毒的
storage ['stɔːrɪdʒ]	*n*.	存储
database ['deɪtəbeɪs]	*n*.	数据库，资料库
encryption [ɪn'krɪpʃən]	*n*.	加密
inadequacy [ɪn'ædɪkwəsɪ]	*n*.	不适当，不充分
static ['stætɪk]	*adj*.	静态的
vulnerability [ˌvʌlnərə'bɪlətɪ]	*n*.	易损性；弱点
reliable [rɪ'laɪəb(ə)l]	*adj*.	可靠的；可信赖的
enhance [ɪn'hæns]	*v*.	提高；加强；增加
dissemination [dɪˌsemɪ'neɪʃn]	*n*.	宣传；散播；传染
centralize ['sentrəlaɪz]	*v*.	使集中；使成为……的中心
backup ['bækʌp]	*v*.	做备份
aggressive [ə'gresɪv]	*adj*.	侵略性的；好斗的
panacea [ˌpænə'siːə]	*n*.	灵丹妙药；万能药

Exercises

Ⅰ. Read the passage above and decide whether the following statements are true (T) or false (F).

( ) 1. Computer network security is to ensure the network storage, security and the transmission of information security.

( ) 2. Powerful integrated intrusion detection system will detect all possible attempts to attack and take corresponding measures.

( ) 3. In the network anti-virus program, we firstly have to reach the goal to put an end to the LAN in the whole virus infection, dissemination and attack.

( ) 4. Network firewall as the first line of defense can protect the internal network fully.

( ) 5. The second line of defense is the main part of a breakthrough to prevent the firewall as well as attacks from within.

Ⅱ. Translate the following paragraph into Chinese.

1. Network security is the security of computer information system in an important aspect. As the opening of Pandora's Box, the computer systems of the Internet,

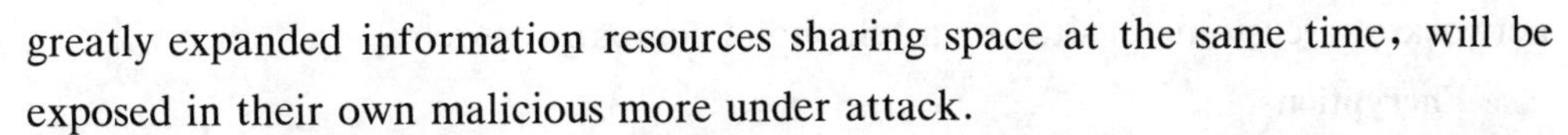

greatly expanded information resources sharing space at the same time, will be exposed in their own malicious more under attack.

2. Using the most advanced vulnerability scanning system scanning workstations, servers, switches and other safety checks regularly, and in accordance with the results to provide the system administrator with detailed and reliable security analysis, in order to enhance the overall level of network security have an important basis.
3. On the whole, network security is a comprehensive issue. It involves technology, management, use and many other aspects, including both its own information system security issues, and physical and logical technical measures. As a kind of technology, it can only solve the problem on the one hand, rather than a panacea.

Further Reading

How to Protect Your Data

Nobody's data is completely safe. But everybody's computer can still be well guarded against would-be attackers. Here's your arsenal:

Password Protection

At minimum, each time they log on, all PC users should type in a password that only they and network administrator know. PC users should avoid picking words, phrases, or numbers that anyone can guess easily, such as a birth date, a child's name, or initials; instead they should use cryptic phrases that combine uppercase and lowercase letters, such as "The moon also RISES." In addition, the system should require all users to change passwords every month or so and should lock out prospective users if they fail to enter the correct password three times in a row.

Virus Checkers

Viruses generally infect local area networks through workstations, so anti-virus software packages that work only on the server aren't enough to prevent infection; ideally, all terminals on the network-personal computers as well as dumb workstations should be protected individually.

Firewall

These are gatekeepers made of hardware and software that protect a computer network by shutting out unauthorized people and letting others go only to the areas they have privileges to use. Firewalls should be installed at every point where the computer system comes in contact with other networks including the Internet, a separate local area network at a customer's site, or a telephone company switch.

Encryption

Even if intruders manage to break through a firewall, the data on a network can be made safe if it's encrypted. Many software packages and network programs offer add-on encryption schemes that encode all data sent on the network with individual applications, almost every encryption package is based on an approach known as public-private key unique to that transmission, receivers use a combination of the sender's public key to and their own private encryption key to unlock the secret code for that message and decipher it. In fact, one scheme, PGP, For Pretty Good Privacy, has almost become a standard since it was offered in 1991 as free software by its creator, Phil Zimmermann.

Audit Trails

Almost all firewalls, encryption programs, and password schemes include an auditing function that records activities on the network. This log—which, ironically, is turned off by many network administrators who don't appreciate its importance—is an excellent way of recording what occurred during an attack by hackers. The audit trail not only highlights points of vulnerability on a network, but also identifies the password and equipment used to invade the system during an inside job. In addition, this auditing log can prevent internal intrusions before they occur when employees know such a trail exists.

Writing

Structure & Format in Agenda Writing

会议议程是整个会议顺序的总体安排，它包括会议期间的活动、任务、理念等。会议议程有助于会议的各项活动按计划进行。在会议前制定合理的议程能提高会议效率，达到事先预计的会议目的。会议议程应包括以下几点：

1）会议开始的时期、时间、地点以及会议持续的时间。

2）参与会议的人员。

3）会议的议题，以及相应的可能的解决方法。

4）确保会议的每个步骤都按照逻辑安排，若涉及之前会议的内容，应在本次会议议程中体现出来。

5）标明每项议题的具体开始时间和持续时间。

6）确保会议议程在会议开始前几天送达与会人员，并附上与会议相关的信息。

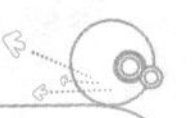

Sample

Date:	20 July, 2012
Time:	9:00 am
Venue:	Conference Room 1, Building 3, Head-office
Participants:	Zhang Fangfang, Yang Lijuan, Miao Chen, Wang Kai
Reading of minutes of previous meeting and approval of the minutes	
—Sales and Market: by Mrs. Zhang Fangfang —Production: by Mrs. Yang Lijuan —Staff Training: by Mr. Miao Chen —Finance and Accounts: by Mr. Wang Kai	
Discussion on the unfinished business (mentioned at the previous meeting):	
Discussion of the new business (presented in the reports):	
Date of next meeting:	
Other business:	
Adjournment (休会):	

Task

A regular monthly meeting of Nortel Marketing is going to be held at Conference Room 800, Sky Hotel, 9:00 am to 11:00 am, September 14, 2012. Three managers (Mr. John, Mr. Green and Mrs. Wang) in this area will make their reports at the meeting. Write an agenda for the meeting by referring to the format given above.

Date:	
Time:	
Venue:	
Participants:	
Reading of minutes of previous meeting and approval of the minutes	
1. 2. 3.	
Discussion on the unfinished business (mentioned at the previous meeting):	
Discussion of the new business (presented in the reports):	
Date of next meeting:	
Other business:	
Adjournment (休会):	

Unit 10 Malfunction and Elimination

Learning Objectives

In this unit you should get familiar with:

- The normal malfunction and the way of elimination;
- The daily ways of discussing the computer malfunction;
- Understanding the principle of elimination.

Speaking

Dialogue 1

Tom: Hey, Jerry. How are you doing today?

Jerry: I am doing fine, and how about you? I think you must be busy these days.

Tom: Yes, I am. Our bank decided to set up a new branch in Zhengzhou, so there are many preparations need to be done.

Jerry: Sound good. So you guys are going to buy a great many new computers for that branch?

Tom: Partly, but I am not responsible for this part, I am just in charge of setting the new cipher machine.

Jerry: Cipher machine, it sounds so mystical.

Tom: Ha-ha. That is because you are not major in computer science.

Jerry: Probably, but it is really a big problem for me to deal with the computer things.

Tom: Really? If you encounter this kind of problems next time feel free to call me, it will be my pleasure to give you a hand.

Jerry: Wow, I appreciate it. It is really good news for me. Oops, it's 6 pm now. I need to go, my son is waiting me for dinner, see you.

Tom: All right. See you.

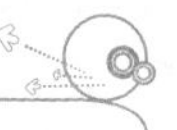

Dialogue 2

Tom : Did you get your new laptop yet?

Jerry: No, I've done some shopping, but it's hard to decide.

Tom : Why's wrong?

Jerry: Well, I've found one that I like quite a bit. It has 15-inch screen with good resolution. But it doesn't have a built-in-CD-ROM. The other has both a floppy drive and CD-ROM built-in. But the screen is small and it's just too heavy.

Tom : I guess you can't have the best of both worlds.

Jerry: I guess not.

Dialogue 3

Tom : What a bad luck!

Jerry: What seems to be the problem?

Tom : My computer just crashed again for the third time this day!

Jerry: What were you doing when it crashed?

Tom : I was just opening up an URL watching a video.

Jerry: I think that might have been a virus.

Tom : Oh, no! I thought it seemed a bit strange.

Jerry: What kind of computer do you have, a Mac or a PC?

Tom : It's a PC. Doesn't everyone have a PC in this office?

Jerry: No, some people have Macs now, too.

Tom : What's the difference?

Jerry: PCs often crash from virus, but it's nearly impossible to get a virus from a Mac.

Tom : I didn't know that.

Jerry: Did you end up losing any of your work?

Tom : Fortunately, I saved my work right before it crashed, so it should be OK.

Jerry: You should probably call the IT department and have them check your computer for virus.

Tom : That's a good idea. I'll call them now. Thanks for your help!

Dialogue 4

Tom : Do you have any experience working with a computer?

Jerry: Yes. I have been a data entry operator for four years.

Tom : What kind of software can you use?

Jerry: I have working knowledge of Windows and Dos. Actually, I'm quite familiar with both Java and C++ Programming Languages.

Tom : Do you have any other computer qualifications?

Jerry: I have an NCRE certificate, Bank 2.

Tom : Do you know how to use a PC to process the management information?

Jerry: I'm sorry to say I'm not familiar with processing management information, but I'm sure I could learn quite quickly. It can't be too difficult, and I've got a quick mind. I can handle any problem you give me.

Dialogue 5

Clerk : Can I help you?

Jerry : OK. I want to buy a computer.

Clerk : What do you want to be able to do with it?

Jerry : Oh, I'm not sure. I don't know much about computers.

Clerk : Do you want to play games or write documents?

Jerry : What's the difference?

Clerk : Well, for game playing, it's better if you have a bigger memory and a better quality video card.

Jerry : I guess Pentium IV 1.6 would be good.

Clerk : OK. What size hard drive do you want?

Jerry : It doesn't need to be too big, just big enough to play games. What about the rest?

Clerk : If you're concerned about game effects, I suggest you buy a higher quality sound card and video card.

Jerry : Effects? Yeah, that's important. Give the best! And I'd like a big screen.

Clerk : The monitor for a big screen will take up more space. What about a flat screen? They're more expensive, however.

Text Reading

Text A

Common Malfunction of Hardware and Software

Hardware and Software Malfunction

Hardware failure is caused by the computer hardware that related to the host system, the memory, keyboard, display and display means, disk drive control means, and power supply components.

Software failure, in the narrow mean, includes the misuse, driver conflicts,

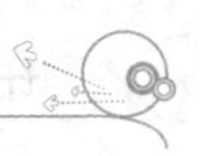

initialization settings wrong and Bug factors of the presence of the software itself. Broadly, software malfunction covers computer as the core, network extension of modern business office systems: systems and applications installed (including new software installed), OA, e-mail handling, viruses and Trojans, hardware and networks, network driver and program failures, application software updates, printer, etc.[1]

Causes of Hardware and Software Malfunction

The poor quality of the hardware itself. Inferior hardware are used to replace original hardware system. It is combined by original system hardware and substandard system hardware that it is very easy to cause disorder or internal short circuit and bad.

Human factors. In actual operation, the operation does not meet the regulated operation of the system that causes the hardware malfunction.

Applicable environmental impacts. Any environmental factor affecting the machine is very huge. Computer hardware operating environment over the allowed limit will seriously affect the performance of your computer, resulting in a hardware failure.[2]

Software failure occurs more frequently, usually a virus infection, security vulnerabilities, Trojan hacking, illegal operation, abnormal shutdown, rogue software nuisance; file corruption, authorized encryption, hardware changes caused by several conditions.

Notes:

[1] Broadly, software malfunction covers computer as the core, network extension of modern business office systems: systems and applications installed (including new software installed), OA, e-mail handling, viruses and Trojans, hardware and networks, network driver and program failures, application software updates, printer, etc. 广义上，软件故障涵盖了以计算机为核心、以网络为外延的现代化商务办公体系：系统及应用程序安装（含新机软件安装）、办公自动化及电子邮件处理、病毒及木马、硬件及网络、网络驱动及程序、应用软件更新、打印机等的故障。

[2] Computer hardware operating environment over the allowed limit will seriously affect the performance of your computer, resulting in a hardware failure.
计算机运行环境超过了硬件允许的极限值都会严重影响计算机的性能，造成硬件故障。

Vocabulary

electronic [ɪˌlek'trɒnɪk]	*adj*.	电子的，电子操纵的
efficiency [ɪ'fɪʃ(ə)nsɪ]	*n*.	效率，效能，功效，实力
hardware ['hɑːdweə]	*n*.	计算机硬件；五金器具
software ['sɒf(t)weə]	*n*.	软件

misuse [mɪs'juːs]	*n*.	滥用；误用；虐待
Initialization [ɪˌnɪʃəlaɪ'zeɪʃən]	*n*.	[计] 初始化；赋初值
malfunction [mæl'fʌŋ(k)ʃ(ə)n]	*n*.	故障；失灵；疾病
substandard [sʌb'stændəd]	*adj*.	不合规格的；标准以下的
operation [ɒpə'reɪʃ(ə)n]	*n*.	操作，运算
impact ['ɪmpækt]	*n*.	影响；效果；碰撞
frequently ['friːkw(ə)ntli]	*adv*.	频繁地，经常地；时常
vulnerability [ˌvʌlnərə'bɪlətɪ]	*n*.	易损性；弱点
abnormal [æb'nɔːml]	*adj*.	反常的，不规则的；
nuisance ['njuːs(ə)ns]	*n*.	损害；麻烦事；讨厌的东西
corruption [kə'rʌpʃ(ə)n]	*n*.	腐败，堕落，贪污
encryption [ɪn'krɪpʃən]	*n*.	加密；加密术

Exercises

Ⅰ. Fill in the blanks with the information given in the text.

1. Inferior hardware are ________ to ________ original hardware system.
2. It is ________ by original system hardware and substandard system hardware that it is very easy to ________ disorder or internal short circuit and bad.
3. Any environmental factor ________ the machine is very huge.
4. Computer hardware operating environment over the allowed limit will seriously affect the performance of your computer, ________ a hardware failure.

Ⅱ. Match each of the following terms to its equivalent(s).

1. initialization	A. 滥用
2. misuse	B. 木马
3. encryption	C. 效率
4. efficiency	D. 易损点
5. vulnerability	E. 加密
6. Trojan	F. 初始化

Ⅲ. Read the passage above and decide whether the following statements are true (T) or false (F).

( ) 1. Hardware failure is caused by the computer hardware that related to the host system, the memory, mouse, etc.

( ) 2. Strictly speaking, software malfunction covers computer as the core, network extension of modern business office systems.

( ) 3. If people don't obey the regulated operation of the system, it could cause the hardware malfunction.

( ) 4. Software failure occurs more frequently than hardware failure.

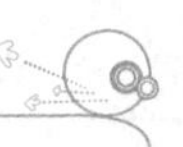

Ⅳ. **Translate the following passage from English into Chinese.**

Instant messaging is like having a real-time conversation with another person or a group of people. When you type and send an instant message, the message is immediately visible to all participants. Unlike e-mail, all participants have to be online (connected to the Internet) and in front of their computers at the same time. Communicating by means of instant messaging is called chatting.

Text B

The Ways of Preventing the Computer from Viruses

You've probably been sick before. In many cases, like the flu, you're sick because of a virus, tiny germs ready to multiply and spread from person-to-person, via handshakes or sneezes. Computer viruses are no different. Instead of germs, they are computer programs. These programs are usually designed by criminals to multiply and spread from computer-to-computer like a disease. If one makes it to your computer, it can erase your files, send emails without your permission or even communicate sensitive info to criminals. Let's take a closer look, because what we call computer viruses can actually be viruses, worms [1] or Trojans[2].

We'll start with viruses. These bugs hitch a ride when something, like a file, is shared between computers. This often happens via attachments sent in email or shared USB drives. Once someone clicks to open the file, the damage is done. The virus is now on that computer, where it starts to multiply and look for chances to hitch a ride to a new computer. Like a sick human, it's sometimes hard to tell when a file has a virus. For this reason, the best defense is anti-virus software. It prevents viruses from getting to your computer and removes them when they are found.

Now, worms are a little scarier. They are programs that spread to computers without humans doing anything. Criminals create worms to spread via computers that are connected in a network. They worm their way from computer-to-computer automatically, whether it's a small office or a global network like the Internet. Usually, the worms find a back door, a way to trick the computer's software into letting them in. Once they're in, they look for the same backdoor in similar computers, wreaking havoc along the way. The best defense is keeping your computer software up to date at work and home. This helps close the doors and prevent problems.

Trojans, our last example, are sneaky bugs. Like the real Trojan horse, they're tricky. If you fall for it, you end up downloading a virus from the Internet. It may appear to be a game or useful software, but hidden inside is a program that can cause problems. For example, these programs can open new backdoors, giving criminals access to your computer and information over the Web. To avoid Trojans, only download software from sites you trust. Just like washing your hands and covering your cough, you have to be aware of what causes problems to avoid them. Keep your computer up to date and get anti-virus software. It will help prevent problems and help you recover. And please, don't click on links, attachments and downloadable files, unless you know they're legit.

Vocabulary

spread [spred]	*v*.	传播，展开，散布
germ [dʒɜːm]	*n*.	细菌，微生物，胚芽
multiply ['mʌltɪplaɪ]	*v*.	乘，使繁殖，使相乘；增加
sensitive ['sensɪtɪv]	*adj*.	灵敏的，易受伤害的
virus ['vaɪrəs]	*n*.	病毒
worm [wɜːm]	*n*.	虫，蠕虫，小人物；使蠕动，使缓慢前进
Trojan ['traʊdʒən]	*n*.	勇士，特洛伊人；木马（电脑）
havoc ['hævək]	*n*.	大破坏，混乱
legit [lɪ'dʒɪt]	*adj*.	合法的，正统的
sneaky ['sniːkɪ]	*adj*.	鬼祟的，卑鄙的

Notes:

[1] worm，称为蠕虫。一般认为蠕虫是一种通过网络传播的主动攻击的恶性计算机病毒，是计算机病毒的子类。网络蠕虫强调自身的主动性和独立性。

[2] Trojan，称为木马病毒，是指通过特定的木马程序来控制另一台计算机。木马这个名字来源于古希腊传说《荷马史诗》中木马计的故事。“Trojan”一词的特洛伊木马本意是特洛伊的，即代指特洛伊木马，也就是木马计的故事。木马程序是目前比较流行的病毒文件，与一般的病毒不同，它不会自我繁殖，也并不刻意地去感染其他文件，它通过伪装自身来吸引用户下载执行，向施种木马者提供打开被种主机的门户，使施种者可以任意毁坏、窃取被种者的文件，甚至远程操控被种主机。

Exercises

Ⅰ. Read the passage above and decide whether the following statements are true (T) or false (F).

( ) 1. These bugs hitch a ride when something, like a file, is shared between computers that often happens via attachments sent in email or shared USB drives.

( ) 2. The best defense is anti-virus software to prevents viruses from getting to

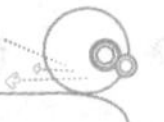

your computer and removes them when they are found.

( ) 3. Computers are sick because of a virus, tiny germs ready to multiply and spread from person-to-person, via handshakes or sneezes.

Ⅱ. **Translate the following sentences into Chinese.**

1. These programs are usually designed by criminals to multiply and spread from computer-to-computer like a disease. If one makes it to your computer, it can erase your files, send emails without your permission or even communicate sensitive info to criminals.
2. Trojans, our last example, are sneaky bugs. Like the real Trojan horse, they're a trick. If you fall for it, you end up downloading a virus from the Internet. It may appear to be a game or useful software, but hidden inside is a program that can cause problems. For example, these programs can open new backdoors, giving criminals access to your computer and information over the Web.

Further Reading

Factors that Affect the Computer Network

There are many factors that affect computer's network security and they are divided into subjective factors and objective factors. Subjective factors generally include man-made malicious attacks, sneak into someone else's computer to find loopholes in the system, the use of non-normal means to steal or destroy data, there are various systems virus, some network resource abuse, loopholes in management, information leaks, etc. Objective factors including the impact of some of the harsh environment, aging equipment, electromagnetic interference and so on computer network security caused a certain degree of threat.

First, information is disclosure and tampering. This refers to the network to upload information from eavesdropping, but does not destroy the network to transmit information. Information has been tampered with is not only refers to the intruder eavesdropping network information, but also made some changes to the information, making information about its authenticity, played a role in misleading information.

Second, there are vulnerabilities of operating system. This is a major threat to network security threats. Since the existence of the computer operation complexity makes network services and network protocols cannot effectively achieve, so in a complex implementation process, resulting in the computer operating system, there are some loopholes and defects.

Third, there is network information openness under the resources sharing. Due to the openness of the network and other information, civilians or government sensitive information is available on the computer network, thus giving criminals can take advantage of the machine, the network elements to provide more detailed information to facilitate their criminal behavior.

Fourth, there are software vulnerabilities. Software vulnerabilities typically include operating systems, network software and services, application software, database, TCP/IP and other vulnerabilities. There are vulnerabilities so that if the computer has a virus attack, it will cause a great threat to network security.

Writing

Notice

通知一般用于上级对下级/组织对成员布置工作/传达事情及召集会议等。通知就其形式而言，可分为书面通知和口头通知两种形式。

书面通知是书面的正式公告或布告，常常张贴在显眼的位置，多用 notice 做标志。为了醒目起见，标志可以全用大写字母，即 NOTICE（但如发出通知的单位以首字母大写形式出现在通知的标志之上，则也要用首字母大写形式的 Notice），并常写在正文上方的正中位置。标志后可以编号，也可以不编号。口头通知用 Announcement 做标志，但通常省略不说。

1. 发出通知的单位及时间

发出通知的人或单位的名称，一般写在标志的上方或在正文后面的右下角；发出通知的时间要写在正文的左下角，也可按书信格式写在正文的右上角。不过，这两项有时可以省略。口头通知不说这两项。

2. 通知的正文

正文要写明所做事情的具体时间、地点、概括性内容（多为书面通知的首句）、出席对象及有关注意事项。布置工作的通知要把工作内容和要求写清楚。正文一般可采用文章式，为了醒目也可采用广告式。广告式要求简明扼要，一个句子可以写成几行，且尽量写在中间，各行的第一个字母一般都大写。

3. 通知的对象

被通知的单位或人一般用第三人称；但如果带有称呼语，则用第二人称表示被通知的对象，口头通知就常用第二人称表示被通知的对象。涉及要求或注意事项时，也常用第二人称表示被通知的对象（祈使句中常常省略）。

4. 通知的文体

书面通知用词贴切，语句简洁，具有书面化；口头通知用词表达要注重口语化。口头通知的开头往往有称呼语（被通知的对象），如“Boys and girls”“Ladies and gentlemen”“Comrades and friends”等；或用提醒听众注意的语句，如“Attention, please!”“Your attention, please!”“May I have your attention, please?”等；且最好有结束语，如“Thank you (for listening).”以示礼貌。

另外，书面通知的写法有两种：一种是以布告形式贴出，把事情通知有关人员，如学生、观众等，通常不用称呼；另一种是以书信的形式，发给有关人员，这种通知写作形式同普通书信，只要写明通知的具体内容即可。通知要求言简意赅、措辞得当、时间及时。

Sample 1

NOTICE

All professors and associate professors are requested to meet in the college conference room on Saturday, August 18, at 2:00 pm to discuss questions of international academic exchanges.

May 14, 2000

Sample 2

Dear Examinee:

As you know, due to unfortunate circumstances, ETS was forced to cancel the scores of the October 1992 TOEFL administration in the People's Republic of China. At that time, you were notified that you would be able to take another TOEFL without charge up through the October 1993 administration. You should be aware that the TOEFL program has a long standing policy of not refunding test fees when administrations are cancelled.

We apologize for any inconvenience that this may cause to you.

Russell Webster

Executive Director

TOEFL Program

Educational Testing Service

Task: Please write a notice according to the following information.

本星期六晚上 7 点到 9 点 0301 班将在演唱厅举行一场晚会，作为 0301 班的班长，你将邀请 0403 班的同学集体来参加，请以此写出一份通知邀请他们参加。

Unit 11 After-Sales Service

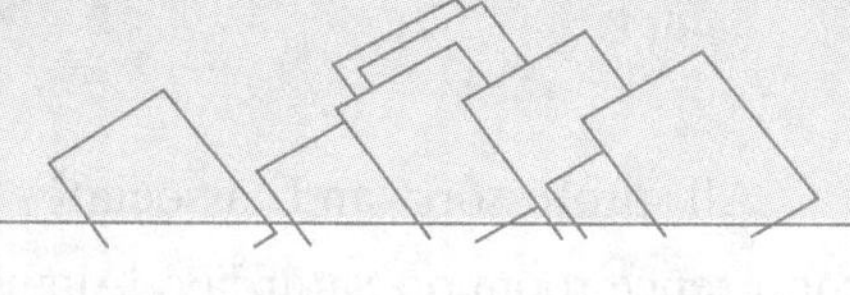

Learning Objectives

In this unit you should get familiar with：

- The main idea of after-sales service；
- The process of after-sales service；
- The ways of complaint；
- The ways of claim.

Speaking

Dialogue 1

A：After-Sales Department of TOC Company，Alex speaking.

B：Hello. I'm calling because there seems to be a problem with my cell phone.

A：Oh，I'm sorry about that. What seems to be the problem?

B：Well，I think I've been constantly hearing cracking noise from it.

A：OK. What would you like me to do?

B：I want you to either have this repaired or exchange it for another one.

Dialogue 2

A：Good morning，After-Sales Department of TOC Company. How can I help you?

B：Oh，good morning. Um，I'm having a bit of problem with a laptop computer I bought recently.

A：What model computer do you have，sir?

B：Oh，it's a TX-220.

A：Could you give me the service code?

B：Oh，I'm not sure if...

A：Do you have the computer close to you? Well，you'll find a number on the underside，somewhere near the battery.

B：OK. Just a minute. Ah，I found it. It's 98D254CF.

Dialogue 3

A: Good morning, TOC After-Sales Department. How can I help you?

B: My printer is not working again! Fix it!

A: Sure, Sir. I feel so sorry and we'll do our best. Do you mind telling me a little more about your printer's problem, please?

B: Every time I tried to print something, I only got blank pages. I don't buy printers to print blank pages, do I!

A: Of course not. I'm so sorry. But this problem may be caused by many reasons, so could I get you through to one of our printer technicians for better service?

B: OK.

A: Well, I'm sure he can help you out.

B: All right.

A: Can I take your telephone number in order to contact you later?

B: All right. 3659570.

A: 3659570. OK.

B: Right, thanks.

Dialogue 4

A: I don't understand how to use this new software.

B: Didn't you read the user's manual?

A: No, I bought a bootleg version, so I don't have a manual.

B: That's why you should only use original software.

Dialogue 5

A: So, why don't you try a computer dating service?

B: First of all, it's too impersonal. Second, how do you know the person on the other end is telling you the truth? They could just be making things up all the time and you would be getting to know someone who doesn't really exist.

A: You're right. I guess computer dating has its bad sides, but it's still a fun way to meet people.

Dialogue 6

A: My phone bill is too expensive.

B: Maybe you should get one of those Internet telephone systems.

A: That's to say you both dial into the Internet so you only have to pay for a local phone call, right?

Text Reading

Text A

HP[1] Returns & Exchanges

Return & Exchange Policy

Return Period

We accept returns and exchanges for up to 21 days after delivery. After the 21-day period, you can get support for any product information from HP Customer Care.[2]

Credits

Credits for returns and exchanges are handled once we make sure of the receipt of the products you buy. Please be aware that this may take your bank 5 to 7 workdays to process into your account.

Refunds

Tax is refunded with returns. Original shipping and handling changes aren't refunded. When some items from your order are returned, the discount for the whole order will no longer apply (e.g., free gifts, buying in quantity). In this case, the discount is subtracted from the price you paid for the returned item. To receive a complete refund, all items from the original purchase must be returned.

HP Customer Care provides easy access to full service and 24/7 support, please:

- Visit Support & Drivers for online support, product information, software and drivers.
- Contact us via these numbers:

For HP products: (800) 4746836.

For Compaq products: (800) 6526672.

To Process a Return or Exchange

To return or exchange an item, you must first get a return material authorization number (RMA) by calling us at (800) 9994747.

1) An e-mail with a FedEx shipping label will be sent to you.

2) Print this label within 5 days.

3) Replace the product in its original packaging and include all items that were originally in the box.

4) Replace any original shipping labels with the new label.

5) Write in bold letters on the package: "RMA (your return material authorization number)".

6) Send it via FedEx.

Notes:

1) We pay for return shipping.

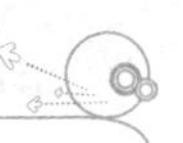

2) You can check the status of your return by viewing order status.

3) For exchanges, we must receive your returned product before we ship the new one.

4) Not following this process may delay your credit or exchange, and you'll have risk of loss and payment of return shipping charges.

Notes:

[1] HP。惠普公司（Hewlett-Packard Development Company，L. P.，HP）总部位于美国加利福尼亚州的帕罗奥多（Palo Alto），是一家全球性的资讯科技公司，主要专注于打印机、数码影像、软件、计算机与资讯服务等业务。惠普（HP）是世界最大的信息科技（IT）公司之一，成立于1939年。惠普下设三大业务集团：信息产品集团、打印及成像系统集团和企业计算及专业服务集团。

[2] We accept returns and exchanges for up to 21 days after delivery. After the 21-day period, you can get support for any product information from HP Customer Care.
我们接受交付后最长21日内的退换请求。超过此21天的期限后，您可从惠普客户服务部获取产品售后支持。

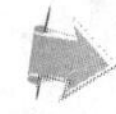

Vocabulary

exchange [ɪks'tʃendʒ]	*v.*	交换，互换；兑换
credit ['kredɪt]	*n.*	信誉，信用
receipt [rɪ'siːt]	*n.*	收据，发票；收入
refund ['rɪfʌnd]	*n.*	资金偿还；偿还数额
return [rɪ'tɜːn]	*v.*	回转，复发
discount [dɪs'kaʊnt]	*v.*	打折扣，减价出售
subtract [səb'trækt]	*v.*	减去；扣除
original [ə'rɪdʒənl]	*adj.*	原始的；独创的
authorization [ɔːθəraɪ'zeɪʃən]	*n.*	授权，批准
replace [rɪ'pleɪs]	*v.*	替换，代替，把……放回原位
package ['pækɪdʒ]	*n.*	包裹；包装袋
label ['leɪbl]	*n.*	标签，标记，符号
bold [bəʊld]	*adj.*	明显的，醒目的；勇敢的

Exercises

Ⅰ. **Read the passage and answer the following questions.**

1. How long is the return period?

2. How long will the bank process credits into the account?

3. What needs to be done before returning and exchanging a product?
4. What should be written on the package?
5. Who will pay for the return shipping?

Ⅱ. **Read the passage above and decide whether the following statements are true (T) or false (F).**

( ) 1. You need to call (800) 4746836 to get a RMA number.
( ) 2. The status of the return can be checked by looking at order status.
( ) 3. The returned product is received after shipping the new item.
( ) 4. You may take risk of loss and payment of return shipping charges if you don't follow this process.

Ⅲ. **Put the following steps of returning or exchanging into correct order. Write numbers before the statements.**

( ) Write RMA on the package in bold letters.
( ) Put the product and all items in its original box.
( ) Get a RMA number by making a call.
( ) Print the FedEx shipping label.
( ) Send it via FedEx.
( ) Receive an e-mail with a FedEx shipping label.
( ) Paste a new label on the product.

Ⅳ. **Fill in each blank with the appropriate form of the words given in brackets.**

original, discount, exchange, receipt, subtract, credit, authorization, replace, package, bold

1. We'll hold a meeting to ____________ experiences.
2. The label should be firmly affixed to the ____________.
3. More new machines will be installed to ____________ the old ones.
4. You cannot take a day off without ____________.
5. The digital computer, in its computation section, can do mainly two things-add or ____________.
6. The translation does not quite correspond to the ____________.
7. After I paid the money, the shop assistant gave me a ____________.
8. Be ____________ in putting things into practice and blazing new trails.

Ⅴ. **Translating.**

A. Translate the following sentences into Chinese.

1. We pay for return shipping.
2. You can check the status of your return by viewing order status.

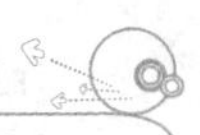

3. For exchanges, we must receive your returned product before we ship the new one.
4. Not following this process may delay your credit or exchange, and you'll have risk of loss and payment of return shipping charges.

B. Translate the following sentences into English.

1. 使用者不应在未授权的情况下安装任何软件。(authorization)
2. 产品不合规格，保证退换。(replace)
3. 现金付款，他们给以九折优待。(discount)
4. 用信用卡付款非常方便。(credit)
5. 你若不满意，我们愿意退款给你。(refund)

Text B

How to Complain?

When people need to complain about a product or poor service, some prefer to complain in writing and others prefer to complain in person. Which way do you prefer? Use specific reasons and examples to support your answer.

When I want to make a complaint about a defective product or poor service, I would rather make my complaint in writing. Writing a complaint allows me to organize my points of argument in a logical manner. If I'm really unhappy with the way I'm being treated, I want to present my reasons clearly. I don't want there to be any confusion about why I'm complaining. I like to list my complaints and then list supporting examples. That's the best way of making sure everyone is clear about what I'm saying.

Putting my complaint in writing also ensures it won't seem too emotional. If you feel that you're been treated badly or taken advantage of, it's easy to lash out. Losing your temper, though, is a sure way to lose your argument. Yelling is very satisfying at the moment, but it only makes the person you're yelling at mad at you. It doesn't get them to agree with you or to offer help.

There's also the issue of the person you're dealing with. If you complaint in person, you have to talk to whoever is there. Chances are that he or she isn't the person responsible for the defective product or the poor service. Often the people who take complaints are not the people in charge, unless you're dealing with a very small business. Yelling at them isn't fair, and doesn't do anything to get a refund or satisfaction for you. You need to reach the person in charge. The best way to do that is in writing.

Writing about your complaint and sending the letter by registered mail also gives you written proof. It's clear that you tried to settle the matter in a reasonable manner within a certain time period. This way, if you need to take further action, you have physical evidence of your actions.

Writing a complaint has the advantages of organization, effectiveness, and fairness. That's why I prefer to write rather than personally present my complaints.

 **Vocabulary**

specific [spə'sɪfɪk]	*adj*.	明确的，具体的
defective [dɪ'fektɪv]	*adj*.	有错误的，有缺陷的
logical ['lɒdʒɪkəl]	*adj*.	逻辑上的，合逻辑的
confusion [kən'fjuːʒən]	*n*.	糊涂；混淆，混同
emotional [ɪ'məʊʃənəl]	*adj*.	强烈情感的
mad [mæd]	*adj*.	疯了的，恼火的，发怒的
registered ['redʒɪstəd]	*adj*.	注册的；登记过的
refund ['rɪfʌnd]	*n*.	归还，退款
evidence ['evɪdəns]	*n*.	证词；证据

 **Phrases and Expressions**

in person	亲自，亲身
yell at	对……喊叫
lash out	猛打，大手笔地花钱
take advantage of	利用

Exercises

Ⅰ. **Read the passage above and decide whether the following statements are true (T) or false (F).**

( ) 1. There may be some confusions when you are complaining.

( ) 2. The best way to complain is in person face to face.

( ) 3. Yelling is very satisfying and it makes the person you're yelling at mad at you.

( ) 4. Often the person who take complaints is not the person in charge.

( ) 5. Writing a complaint has the advantages of organization, effectiveness, and fairness.

Ⅱ. **Choose from the following words and expressions and fill in the appropriate box.**

specific, defective, confusion, registered, evidence, yell at, mad, refund, take advantage of, lash out

1. We haven't fixed ________ date for our meeting.

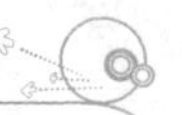

2. If the goods prove ___________, the customer has the right to compensation.
3. His unexpected arrival threw us into total ___________.
4. Never walk behind a horse in case it ___________.
5. The noise outside the building nearly drove me ___________.
6. Article 10 provides that all businesses must be ___________.
7. They demand _______ on unsatisfactory goods.
8. We should _______ this chance sufficiency.

Further Reading

Customer Loyalty

Customer loyalty describes the tendency of a customer to choose one business or product over another for a particular need. In the packaged goods industry, customers may be described as being "brand loyal" because they tend to choose a certain brand of goods more often than others. Customer loyalty becomes evident when choices are made and actions taken by customers. Customers may express high satisfaction levels with a company in a survey, but satisfaction does not equal loyalty, that is demonstrated by the actions of the customer. Customers can be very satisfied and still not be loyal.

Customer loyalty has become a catchall term for the end result of many marketing approaches. Increased customer loyalty is the end result, the desired benefit of these well-managed customer retention programs. Customers who are targeted by a retention program demonstrate higher loyalty to a business. All customer retention programs rely on communicating with customers, giving them encouragement to remain active and choosing to do business with a company. It takes a lot less money to increase your retention of current customers than to find new ones.

There are four factors that will greatly affect your ability to build a loyal customer base: products highly differentiated from those of the competition; higher-end products where price is not the primary buying factor; products with a high service component and multiple products for the same customer.

When customers are not happy with your business they usually will not complain to you, instead, they will probably complain to just about anyone else they know-and take their business to your competition next time. That is why an increasing number of businesses are making follow-up calls or mailing satisfaction questionnaires after the sale is made. They find that if they promptly follow up and resolve a customer's complaint, the customer might be even more likely to do business than the average customer who did not have a complaint.

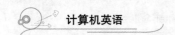

Companies should not try to manage loyal customers; long-standing relationships arise from trust gained over many transactions, and they are sustained by customers' belief that the company wishes to keep them around rather than drive them away.

In many business situations, the customer will have many more interactions after the sale with technical, service, or customer support people than they did with the sales people. So if you are serious about retaining customers or getting referrals, these interactions are the ones that are really going to matter. They really should be handled with the same attention and focus that sales calls get because in a way they are sales calls for repeat business.

Building customer loyalty will be a lot easier if you have a loyal workforce. It is especially important for you to retain those employees who interact with customers such as sales people, technical support, and customer-service people. Many companies give a lot of attention to retaining sales people but little to support people. Clearly, customer loyalty is too central to companies' fortunes to be left to the marketing departments alone. And technology is also important in determining retention or customer disaffection.

Contact with current customers is a good way to build their loyalty. The more the customer sees someone from your firm, the more likely you will get the next order. The more they know about you, the more they see you as someone out to help them. The more they know about your accomplishments, the more loyal customers they will be.

Customer loyalty seems like a quaint notion in the Internet age, when customers can search out lower prices and defect to competitors with a mouse-click. Yet in the faceless online market, customers yearn for trustworthiness more than ever. Give it to them and they are yours forever.

Writing

Complaint and Claim

抱怨、索赔信函的目的是获取更好的服务，对已出现的问题求得尽快、妥善的解决。它通常是买方因为对收到的货物不满而书写的。例如：货物未按时到达；货物虽抵达，但与订单所载不同；货物有损毁现象；货物数量短缺或多余；货物质量与样品不符；服务不合理以及收费过高等，这些均是抱怨、索赔的正当理由。

书写抱怨、索赔信函时，不妨开门见山给出原先双方同意的条件，然后列举事实以表示有何不满，以及为什么不满，最后提出解决的方法。内容应明确、清楚、有理，语气要简洁、坚决。避免使用愤怒和使对方过于难堪的措辞（除非你所抱怨的问题反复发生，且多次投诉而毫无结果）。

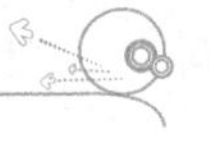

Sample 1

Complaint for Delay in Shipment

Dear Sirs,

The furniture we ordered from you should have reached us a week ago. Needless to say, the delay in delivery has put us to great inconvenience. It is therefore imperative that you dispatch them immediately. Otherwise we shall be obliged to cancel the order and obtain the goods elsewhere.

Please look into the matter as one of urgency and let us have your reply as early as possible.

Yours faithfully,

亲爱的先生：

我方从贵方订购的家具应于一星期前收到。不用说，发货的延迟给我方带来了很大的不便。因此，贵方必须立即发货，否则我们将被迫取消订单，到另处订货。

请紧急处理此事，并尽快告知结果。

Sample 2

Claim for Inferior Quality

Dear Sirs,

With reference to our order No. 315, we are compelled to express our strong complaints for the inferior quality. Compared to the sample No. 169, the arrived self-adhesive correction tape is very transparent and does not satisfactorily cover the error being corrected.

We shall be glad to have your explanation of this inferiority in the quality, and also to know what you purpose to do in this matter.

yours sincerely,

亲爱的先生们：

关于我方第 315 号订单，由于到货质量低劣，我方被迫表达强烈的不满。同 169 号样品比较，此次运到的修改带稀薄，无法满意地覆盖在要被改正的错误上。我方希望贵方对这一情况加以解释，并告知贵方对此事的处理方法。

处理顾客投诉和抱怨有以下技巧。

a）Customer is always right.

b）Response promptly.

c）Remain polite if complaint is unreasonable.

d）Accept the complaint if you are to blame.

e）Never lay the blame on your colleagues or your subordinates.

f）Thank the customer for making a complaint.

Sample 3

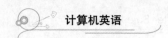

Dealing with a Justified Complaint

Dear Mr. Robinson,

Your complaint about give brief details has been passed to me for attention. I am very sorry that you have been inconvenienced by our failure to provide the level of services to which you are entitled.

I have made enquires and found that the problem was caused by give brief details. I can assure you that we have taken steps to make sure that this kinds of situation cannot arise again.

Please accept our apology and appreciation of your highly valued custom. Please contact me personally if you experience any problems with our services in the future.

Yours sincerely,

Sample 4

Dealing with an Unjustified Complaint

Dear Mr. Robinson,

Thank you for your letter of (date), concerning (brief details of complaint).

I have thoroughly investigated this complaint and have interviewed all the staff involved. I can, however, find no evidence to support the claims that you made. Indeed, I am satisfied that our staff acted properly and with due courtesy and efficiency.

I hope that you will accept my regret that you feel we were remiss in our dealing with you. If, however, you can give me any evidence to support your claims, I shall be pleased to review the situation. If you would like to discuss the matter in person, please contact me directly.

Yours sincerely,

Task: Compose a complaint in the following situations.

亲爱的先生们：对于我方12月20日订购的瓷器，非常感谢贵方及时发货。然而，今早贵方承运人送来360箱货物，而我方只订购了320箱。很不幸，我方目前的需求已完全饱和，无法接受多余的货物。请传真通知我方如何处理。

Unit 12 Prospects of IT

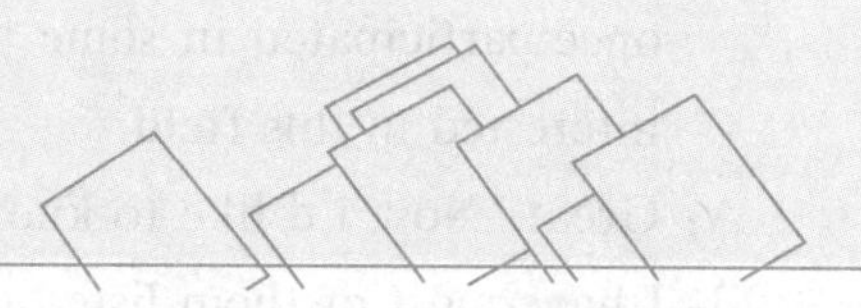

Learning Objectives

In this unit you should get familiar with:

- The skills of interview;
- The relationship of coworkers;
- The developing trend of IT;
- The career trends;
- The writing of resume.

Speaking

Dialogue 1

A: Computers are really spreading quickly, I just found the web site of the high school I attended.

B: Lots of schools have their own web sites.

A: That's true, but when I was in high school, we just had a few computers, and now it looks like the whole school is computerized.

Dialogue 2

A: So, tell me again what this new job you're taking is?

B: I'll be doing web design.

A: Web design? That sounds like work for a spider?

B: I'm talking about designing pages for the World Wide Web—the Internet.

Dialogue 3

A: Your name, please?

B: Jimmy Smith.

A: Nice to meet you, Mr. Smith.

B: I've been looking forward to this interview. These are my personal documents about my education and experience.

A：OK. Can you briefly introduce yourself?

B：I majored in computer in college and I graduated last year with good scores. I once participated in some projects as a tester or a programmer. And I'm very interested in this field.

A：Great. Now I'd like to know more about the projects you just mentioned.

B：I have most of them listed in my resume.

Dialogue 4

A：Well, Mr. Smith. I've looked over your resume. It is impressive. And Diana said you did very well in the interview.

B：Thanks.

A：However, we do have several other applicants to finish interviewing before we can make a final decision. We'll call you by Friday, if it's all right.

B：That's no problem.

A：Do you have any other questions?

B：Well, could you explain a bit about the welfare of the company?

A：Oh, yes. We have an excellent retirement plan and medical insurance for every employee.

B：Great. Thanks so much for your time.

A：We'll contact you soon. Thanks for coming in.

Dialogue 5

A：Hey, Jimmy, I saw this ad in the paper. You should take a look.

B：What is it?

A：It's for a job vacancy. It looks perfect for you.

B：Let me see. "Wanted：technical support engineer. Must have good knowledge of Windows and Linux, deep understanding of testing and troubleshooting, the ability to program, good organizational skills. Related working experience is preferred. Please contact Diana March." Oh, I don't know...

A：Come on, what have you got to lose?

B：What about my resume ?

A：Here, I'll help you type one.

B：Thanks, Helen. You are so kind. I'll call Diana now.

Dialogue 6

A：I'm really fed up with Harry! He's the biggest airhead I've ever met. He always

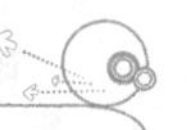

makes careless mistakes, and he's a pain to work with.

B: You shouldn't be so negative. You'll always have some coworkers that are harder to work with than others. But if you are negative and start name-calling in the office, it will make a bad working environment for everybody.

A: You only say that because you don't have to work with him. The people in your department seem capable and nice to be around. Take Mary for example. She's smart and enthusiastic. I've never met anyone as cheery as she is.

B: Everybody has their strengths and weaknesses. Even Larry. He might be a pain to be around, but he's also very good at staying in budget on projects. Mary, on the other hand, spends our project money like there's no tomorrow. Also, she's never willing to stay a little later at the office. She always leave at 5:00 pm sharp.

A: Isn't there anyone in the office that is a perfect coworker? What about Bob? Everybody loves Bob. Even though he's fresh out the college and still a bit green, he is a great co-worker.

B: You are right. He's a hard worker, easy to get along with, honest, and he never steals the credit on projects. The only thing he's lacking in is experience.

A: Maybe that's why he's so nice!

Text Reading

Text A

The Merging of Man and Computer in 21st Century

Education in 2020: Toward Computer Aided Self-Learning

Students typically have a computer of their own, a thin, tablet-like device weighing under a pound, with a very-high-resolution display suitable for reading. Learning materials are accessed through wireless communication. The traditional mode of a human teacher instructing a group of children is still prevalent, but schools increasingly rely on software approaches. Teachers primarily attend to issues of motivation, psychological well-being, and socialization.

Business in 2020: Virtual Commerce[1] Grows

At least half of all transactions are conducted online. There is a strong trend towards the geographic separation of work groups. People are successfully working together despite living and working in different places. The average household has more than two computers, most of which are embedded in appliances and built-in communication systems. Household robots have emerged but are not yet fully accepted.

Health Care in 2020: Electronic Paramedics

Telemedicine[2] is widely used: Physicians examine patients using visual, auditory, andhaptic (tactile) examination from a distance. Computer-based pattern recognition is routinely used to interpret imaging data and other diagnostic procedures. Life-time patient records are maintained in computer databases.

The Arts in 2020: Media-Enhanced Expression

The computer screen is the medium of choice for visual art, which is becoming collaboration between human artists and their intelligent are software. Technology allows nonmusicians to create music, such as cybernetic music-creation systems and software to create music from a person's brain waves. Writers use voice-activated word processing and style-improvement software.

Notes:

[1] Virtual commerce 意为“虚拟商业”，是基于互联网平台，聚集消费者与商家的虚拟商业活动，主要通过互联网进行商家与消费者之间的商品交易。

[2] Telemedicine 指远程医疗，是指以计算机技术，遥感、遥测、遥控技术为依托，充分发挥大医院或专科医疗中心的医疗技术和医疗设备优势，对医疗条件较差的边远地区、海岛或舰船上的伤病员进行远距离诊断、治疗和咨询。

Vocabulary

wireless ['waɪəlɪs]	*adj*.	不用电线的
prevalent ['prevələnt]	*adj*.	流行的，盛行的
motivation [ˌməutɪ'veɪʃən]	*n*.	动机；动力
psychological [ˌsaɪkə'lɒdʒɪk(ə)l]	*adj*.	心理的，精神上的
embed [ɪm'bed]	*v*.	把……嵌入，栽种
auditory ['ɔːdɪtərɪ]	*adj*.	听觉的，听觉器官的
haptic ['hæptɪk]	*adj*.	触觉的
diagnostic [daɪəg'nɒstɪk]	*adj*.	诊断的
collaboration [kə, læbə'reɪʃən]	*n*.	合作
cybernetic [ˌsaɪbə'netɪks]	*adj*.	控制论的

Phrases and Expressions

voice-activated word processing	声控文字处理系统
style-improvement software	文体改进软件

Exercises

Ⅰ. Fill in each blank with the appropriate form of the words given in brackets.

wireless, prevalent, motivation, psychological, embed, auditory, hepatic, diagnostic, collaboration, cybernetic

1. The stronger the ________, the more quickly a person will learn a foreign language.
2. Can you get a ________ access point?
3. The habit of traveling by airplane is becoming more ____________.
4. How to ____________ this virtual tour in my web page?
5. The two companies are working in close ____________ each other.
6. The doctor will conduct a complete ____________ evaluation.
7. Work also provides ____________ well-being.
8. Finally he overcame the ____________ difficulties by three years' efforts.

Ⅱ. **Translating.**

A. Translate the following sentences into Chinese.

1. The traditional mode of a human teacher instructing a group of children is still prevalent, but schools increasingly rely on software approaches.
2. Technology allows nonmusicians to create music, such as cybernetic music-creation systems and software to create music from a person's brain waves.

B. Translate the following sentences into English with the given words.

1. 未来，可在报纸版面中嵌入视频画面，使其成为真正意义上的多媒体报纸。(embed)
2. 她和她的一个学生合写了这本书。(collaboration)
3. 做这个决定的原因是为了改善我们对顾客的服务。(motivation)

Text B

Videonet

A couple of years ago, only a handful of Internet sites existed for publishing videos uploaded by users. Today there are more than 225 such sites, providing the infrastructure to deliver videos created by amateurs and professionals alike.

What's next? As video content—the distribution of which has been historically controlled by a few broadcast networks and cable companies—meets the decentralized, user-centric World Wide Web, are we seeing the dawn of a new medium, a "videonet" that will redefine the media landscape? Experts predicted that virtually every organization marketing to consumers—from TV stations and sports teams to soft drink and detergent makers—will rapidly develop a video presence on the Internet. And it may not stop there. If video publishing grows at a rate similar to that of websites and blogs in recent years, what does it mean for traditional broadcasters, businesses, and users alike?

For the time being, amateur productions dominate. Most sites provide free uploading and are attempting to generate revenue through advertising or distributing commercial video content from traditional broadcast networks, movie studios, and other content partners.

The user-generated content runs the gamut from thought-provoking interviews to comedic parodies to brutal scenes of street fighting, as amateur videographers work in areas that are both accessible and interesting to them, largely eschewing mainstream content and production styles. Among the most popular types are genre videos (such as snowboarding or car racing), parodies and satires, commentaries, interviews, mini TV shows, and documentary film shorts.

Given the size of the video enthusiast market, the key questions are: "How do you build a business around this? How do you create products for this media and audience?"

Media companies are leading the charge. Every traditional news organization is creating a video presence on the web, with newspapers every bit as active as television broadcasters. None can afford to ignore market data that shows their audiences are declining based on the growth of the Internet as a news source. A recent study showed that more than 50 million people in the US use the Internet daily as their primary news source.

Specialized broadcasters are not far behind; in fact, most narrow-interest cable TV stations are near the end of their era, predicted former Internet entrepreneur Jonathan Taplin, now at the University of Southern California's Annenberg Center for Communication. "Today we have about 440 niche cable stations; in 5 years, 300 of them will disappear and the rest will go to the Internet. Discovery, for example, is launching new, narrow channels on the Internet because it can't afford to do them elsewhere."

Vocabulary

infrastructure ['ɪnfrə'strʌktʃə]	*n.*	基础结构，基础设施
decentralize [di'sentrəlaɪz]	*v.*	下放权力
detergent [dɪ'tɜːdʒənt]	*n.*	清洁剂
blog [blɒg]	*n.*	网络随笔，博客
dominate ['dɒmɪneɪt]	*v.*	支配，占优势
revenue ['revənjuː]	*n.*	财政收入
gamut ['gæmət]	*n.*	全音阶
provoking [prə'vɒkɪŋ]	*adj.*	刺激人的
parody ['pærədɪ]	*n.*	拙劣的模仿
brutal ['bruːt(ə)l]	*adj.*	野蛮的
accessible [ək'sesəbl]	*adj.*	可进入的
eschew [ɪs'tʃuː]	*v.*	避开，远避
niche [nɪʃ]	*n.*	适当的位置

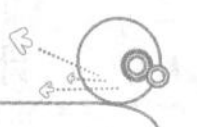

Exercises

Ⅰ. **Read the passage above and decide whether the statements are true (T) or false (F).**

( ) 1. There are few Internet sites for publishing videos in the world today.

( ) 2. Not only can professional users with high computer skills make videos, but also common people can.

( ) 3. Videonet is completely another world different from the ordinary Internet.

( ) 4. Videonet will have a deep effect on traditional broadcasters, businesses, and users.

( ) 5. At present, most sites make large quantity of money through providing video productions.

( ) 6. Most videonet content is sports videos, parodies and satires, commentaries, interviews, mini TV shows, and documentary film shorts.

Ⅱ. **Complete the following sentences by translating the Chinese in the brackets.**

1. These sites ________ (提供一个发布视频作品的平台) created by amateurs and professional alike.
2. ________ (视频内容的发布) has been controlled by few broadcast networks and cable companies.
3. For the time being, ________ (业余爱好者的作品占据主导地位).
4. Every ________ (面向消费者的机构) will rapidly develop a video presence on the Internet.
5. Most sites are ________ (通过……盈利) advertising or distributing commercial video content from traditional broadcast networks, movie studios, and other content partners.

Further Reading

Future of Computers

Every year, IBM Corporation chooses five new technologies it believes will change the world within the next five years. The IBM list is called "Five in Five". The company says it considers its own research and the new directions of society and business when identifying the technologies.

This year, the list describes some future devices that will extend our five senses. Imagine looking for clothes online and touching your computer or smartphone to feel the cloth. IBM Vice President Bernie Meyerson predicts that technology could be available in the next five years.

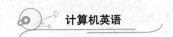

"You're talking about almost reinventing the way computers operate and how you interact with them as humans."

Touch is just one of the senses that computers will help to extend. IBM says smart machines will soon be able to listen to the environment and give us information about the sounds they hear. For example, Bernie Meyerson says an advanced speech recognition system will tell new parents why their baby is crying.

"From the sound the baby is creating, that particular frequency in the voice of the child, you know the difference between a child for instance who is sick as opposed to a child who is just lonely. That kind of understanding would be great for parents. This kind of thing is not possible today, but with a sophisticated enough system, it actually is possible."

Smart machines will also help identify medical conditions. If you sneeze on your computer or cell phone, the machine will study thousands of molecules in your breath. Then it can tell you whether you need to see a doctor.

"It can give you an alarm and say: 'Hey, you may not feel sick yet, but you have an infection, which you must go see your doctor immediately.'"

In the near future, built-in cameras in our personal computers will be able to examine and name colors and recognize images. Mr. Meyerson says IBM scientists are also developing a computer system that can examine and combine food molecules to create the most popular flavors and smells.

"It'll start to be able to recommend to you foods you'll love the taste of, but it can also keep track of the caloric limits, whether you have limits on fat or cholesterol that you can eat. So it strikes this almost ideal balance between the best possible taste and the best possible nutritional outcome."

Mark Maloof is a computer science professor at Georgetown University. He says he hopes the progress that IBM is predicting will lead more students to create future inventions.

"It's going to be very exciting to see what young people do with the increased availability of mobile platforms and networking and computing power."

Professor Maloof says advances in computer technology will make what now seems like science fiction a part of our everyday lives.

Writing

Resume

简历（resume），就是对个人学历、经历、特长、爱好及其他有关情况所做的简明扼要的书面介绍。简历是有针对性的自我介绍的一种规范化、逻辑化的书面表达。标准的求

职简历主要由四个基本内容组成：

1）基本情况：姓名、性别、出生日期、婚姻状况和联系方式等。

2）教育背景：按时间顺序列出初中至最高学历的学校、专业和主要课程，以及所参加的各种专业知识和技能培训。

3）工作经历：时间顺序列出参加工作至今所有的就业记录，包括公司/单位名称、职务、就任及离任时间，应该突出所任每个职位的职责、工作性质等，此为求职简历的精髓部分。

4）其他：个人特长及爱好、其他技能、专业团体和证明人等。

Sample 1

RESUME

Personal Information

Family Name：Wang　　Given Name：Bin

Date of Birth：July 12，1990　　Birth Place：Beijing

Gender：Male　　Marital Status：Unmarried

Telephone：(010) 62771234　　E-mail：career@sohu. com

Objectives

To obtain a challenging position as a LAN designer or software tester.

Education Background

2008 – August 2012　　Dept. of Automation，Tsinghua University，B. E.

Work Experience

Nov. 2012 – present　　CCIDE Inc，as a director of software development and web publishing.

Computer Abilities

Skilled in use of MS Frontpage，Win XP，Java，HTML，JavaScript，C++ and SQL

English Skills

An overall command of English speaking，listening，reading and writing

Passed CET-4

Scholarships and Awards

2009. 03 First-class scholarship for graduate

2011. 07 Metal machining practice award

Qualifications

Aggressive，independent and be able to work under a dynamic environment.

Have coordination skills，teamwork spirit.

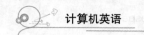

Task: Read the resume and put the following headings into the proper position.

PERSONALITY	INTEREST	PERSONAL INFORMATION
WORK EXPERIENCE	EDUCATION	OBJECTIVE SKILLS

RESUME

________1________

Name: Lin Xinyi

Gender: female

Date of birth: 25 March 1986

Marital status: single

Address: 99 Aiguo Road, Shenzhen

Tel: 130056999

E-mail: linxinyi@hotmail.com

________2________

Administrative Assistant

________3________

2004 – 2007 Jiangsu Polytechnic

Majoring in Business English

Main courses taken: Business English, Practice of International Trade, Human Resources Management, Business Interpretation, E-commerce

________4________

07/2007 – present Secretary to General Manager, Shenzhen Star Electronics Co., Ltd.

Main duties: receiving visitors, arranging meetings and appointments

________5________

Fluent English, conversational Japanese

Working knowledge of Microsoft Word, Excel, and PowerPoint

Driving license

________6________

Outgoing, creative, cooperative, hardworking

________7________

Tennis, music, traveling

Appendix A
参考译文

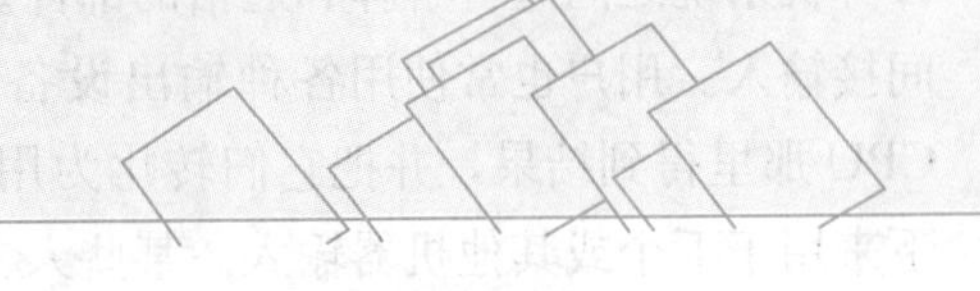

第一单元　计算机概述

Text A

计算机的发展

计算机的历史分为几个发展阶段，所以就有好几代计算机。

第一台数字电子计算机 1946 年诞生于美国，它的基本元件是电子管。[1] 20 世纪 50 年代，还制造了另外几台。它们是第一代计算机，体积大、重量重、价格贵、速度慢且消耗的能量比现在计算机消耗的能量要多得多，但它们依然为计算机科学做出了重大的贡献，如程序存储和随机访问等。[2] 它们确定了现代电子计算机的基本模型。

晶体管的发明不仅给人们带来了袖珍无线电，也促进了第二代计算机的产生。第二代计算机变得体积小、重量轻且价格便宜但还没能轻便和便宜到进入家庭。

20 世纪 60 年代，集成电路问世。集成电路意味着巨大的复杂电路和上百万个元件可以做在一小片半导体的芯片上。[3] 它被引进第三代计算机输出中。第三代计算机的代表是 IBM 公司的 360 系列计算机。大规模集成电路使数字计算机如此普及，以至于每个中等收入家庭都可随便买它们。这就是为什么你在哪里都可见到 PC 机的原因。

计算机已经在很多方面改变了我们的生活。计算机为人们提供了史无前例的获取信息的渠道，改变了社会的隐私状况。[4] 信用卡、银行和电话公司记录着用户的商务活动情况。提供在线搜索、地图和电子邮件等免费服务的互联网公司保存用户输入的信息，并通过机器里的独特识别号码对信息进行分类。

随着科学和技术的发展，生物计算机和量子计算机将在不久的将来面世，新一代计算机将诞生。

Text B

计算机系统

一个系统即是指一组相关部件协作运行。一个数字计算机系统包括 4 个部件：输入设备，输出设备，存储器和中央处理器（CPU）。CPU 能在指令的指导下很快地接受、存储和处理数据或符号并产生结果。在 CPU 处理了输入设备之后，输出设备就给出了用户所需的信息。

用户很少接触 CPU，但都用过输入设备。在 PC 系统中，常接触的是键盘、鼠标、输入笔、触摸屏、话筒和其他直接输入设备。尽管它们之间存在差别，但它们都是在用户和计算机系统之间进行解释和通信的部件。存储部件、软盘驱动设备和硬盘驱动设备常用于间接输入。用户也常使用各种输出设备，如显示器、打印机和绘图仪，这些输出设备从 CPU 那里得到结果，并把它们转化为用户可理解的形式。软盘和硬盘常把这些结果记录下来用于下个或其他机器输入。某些人喜欢把输入输出设备、显示器、打印机和绘图仪等统称为外围设备。

按传统方法，人们将计算机分为两类：数字计算机和模拟计算机。数字计算机处理数字和符号；模拟计算机处理有关量方面的问题，如电流或电压。但现在前者将越来越强大并取代后者。

第二单元　产品设计与操作

Text A

乔纳森·艾维[1]：设计苹果的幕后功臣

苹果公司的产品研发项目常常被隐藏在公众和内部员工的视线之外，其通过对研发过程的严格保密展现了产品的完美无缺。[2]与其下一代电子产品极为相似的是，苹果公司的工业设计师乔纳森·艾维也同样被隐藏在公众的视线之外。

艾维名下有 300 多项设计专利，他的创新技术和方法为“后个人计算机时代”[3]的到来铺平了道路。他对细节一丝不苟的态度推动了消费电子业的发展步伐。与他的好朋友史蒂夫·乔布斯一样，艾维在苹果公司的复兴中发挥了极其重要的作用，但尽管苹果公司因他参与设计的创新产品而举世闻名，他的功劳却在很大程度上被人们忽视。

曾经有三年，艾维都在身不由已地设计牛顿设备和打印机纸盘，其中大多数产品从未上市销售。他甚至设计出了第一台平板电脑，但其昂贵的价格立即让消费者对其避而远之。由于看不到施展才华的机会，艾维开始想办法离开美国。

但在史蒂夫·乔布斯回归苹果后，艾维的悲观情绪很快转变为满怀希望的乐观。在一次对公司园区的巡视中，乔布斯无意中看到了艾维的设计模型。于是，这个如今已是偶像般的人物——消费电子工业界最大公司之一的总裁——要求提拔艾维，并交给他设计 iMac[4]电脑的任务，打算以此缔造苹果的未来。

iMac 的外壳线条优美，呈半透明状，这一特质使得它在同类计算机中独树一帜，一经推出就立刻大受欢迎。尽管它可能没有完全达到乔布斯的期望，在鼠标设计和 USB 技术选用方面令乔布斯不太满意，但它却正好符合消费者的需求，是一台价格便宜的阴极射线管[5]计算机。iMac 的推出挽救了苹果，使其免于在财政危机中垮掉。

艾维设计的下一个产品是 iPod，一款于 2001 年推出的多媒体播放器，它既拥有简洁的外观，又拥有方便的操作和数据传输功能，因此获得了巨大成功。iPod 系列为苹果重新注入了活力，使其成为多媒体行业的领军企业。在此胜利的基础上，艾维又先后设计出了 iPhone 手机和 iPad 平板电脑。

这位广受赞誉的工业设计师拥有独特的理念，即以用户为中心，而不是以产品为中心。在推出 iPad 的时候，这位深受好评的设计师和创新者分享了自己最为重要的设计原则：让用户定义产品，而不是让产品定义用户。这一观点永远地改变了整个技术领域。

在目前平板电脑市场的开创方面，艾维发挥了重要作用。在简单得令人难以置信的 iPad 问世之前，平板电脑市场并不具备现在的形态。通过创造出具有巧克力棒外形的 iPod，他改变了数字音乐产业。通过最近设计的 Macbook 模型和 iPad，他将单一铝板的用途发挥到了极致，使其成为苹果公司的标志性材料。

2010 年，艾维因设计苹果产品而被《财富》杂志誉为“世界上最聪明的设计师”。因设计 iPhone、iPad、MacBook、iMac 和 iPod，被授予多项业界殊荣。至于他最至高无上的一项成就，不是某个具体的奖项，而是英国女王在 2005 年所表达的她对 iPod 的喜欢。

尽管我们无法完全了解艾维的工作过程，但请放心，他仍在实验室里和他的同事们一起努力，为计算机工业的发展铺平道路。虽然我们永远不会了解艾维是如何工作的，但他的工作无疑是“神奇的”。

Text B

操作系统

操作系统是一种程序，起到计算机的用户与硬件之间的接口作用。其目的是提供一种用户能执行程序的环境。因此，操作系统的主要目标就是使计算机系统使用方便。其第二个目标就是要在以一种有效的方式来使用计算机硬盘。

一个操作系统与一个政府类似。计算机系统的基本资源由硬件、软件与数据来提供。操作系统为计算机提供正确使用这些资源的方法。像一个政府一样，操作系统本身执行不了任何有用的功能。它只不过是提供一个环境，在该环境中其他程序能发挥作用。

我们可以把操作系统看作是一个资源分配器。计算机系统有很多的资源（硬件和软件）用来解决一个问题：CPU 的时间，存储内存大小，文件存储空间，输入/输出（I/O）设备等。操作系统就像管理这些资源的经理，为这些资源分配特定的程序和用户作为它们任务的需要。因为在许多资源请求方面可能存在冲突，操作系统必须决定给哪些请求分配资源以便计算机系统能合理而有效地运行。

操作系统与计算机结构彼此有着很大的影响。为了方便硬件的使用，操作系统应运而生。正如操作系统被设计和应用那样，很明显，改变硬件的设计能简化操作系统。通过对历史的简短回顾，我们注意到新的硬件特性的引入自然而然地导致许多操作系统问题的解决。

第三单元 建立商务关系

Text A

苹果的宠儿

—iPad Air

苹果公司赶在圣诞节购物旺季之前推出了 iPad 系列的新品，削减了 Mac 计算机的价格。这些措施都是苹果公司面对其在平板电脑市场上缩小的市场份额，和来自其他平板制造商的日益剧烈的竞争压力所做出的努力。[1]

在周二于美国加利福尼亚州的旧金山举行的新品发布会上，苹果公司展示了一款更新、更纤巧、更轻薄的平板电脑“iPad Air”[2]，还展示了许多 Mac 计算机系列的新款。相比之前净重约 1.4 磅（约 0.64 千克）的旧款 iPad，iPad Air 的重量只有约 1 磅（约 0.45 千克）左右。苹果公司市场总监菲尔·席勒说，iPad Air 的速度“快得令人尖叫”。据他说，它比 2010 年发布的 iPad 要快上 8 倍。

iPad Air 将于 11 月 1 日起发售，初始售价为 499 美元，内备有 16G 的内存卡。[3]苹果公司还计划逐步淘汰掉第三代和第四代的 iPad，而 2011 年推出的第二代 iPad 继续以 399 美元的价格出售。同时，一款新的迷你 iPad 将在 11 月下旬面世，初始售价为 399 美元，内备 16G 的内存卡。这款迷你 iPad 配有画质清晰细致的视网膜屏幕，以及和 iPad Air 上配备的相同的 64 位的芯片。[4]

苹果公司还更新了计算机系列。苹果公司市场总监菲尔·席勒称，新款的 13 英寸 Mac Book[5]配备有视网膜屏幕，而且更纤薄更轻盈。同时，这款笔记本式计算机的电池可以持续使用 9 小时，用户可以用一块电池的电量看完《蝙蝠侠：黑暗骑士》三部曲[6]。而且新款笔记本的价格也更低，相比之前的 1499 美元售价，新款笔记本的售价仅为 1299 美元。

苹果公司还推出了一款尺寸稍大的 Mac Book Pro，15 英寸显示器 256G 内存，售价为 1999 美元，比之前款式的 2199 美元售价也有所下降。

另一款以苹果公司称之为“高级用户”的群体为目标群体的高端笔记本式计算机 The Mac Pro，将于 12 月上市，价格为 2999 美元。

苹果公司还称其最新研发的计算机操作系统 Mavericks 将免费供用户使用。

Text B

关于笔记本式计算机的进一步咨询

亲爱的露西：

在和贵公司销售经理李先生见面及打电话谈论购买意向后，我们对贵方所推荐的 Mac Book Pro 比较感兴趣。

您知道，我们是以职业培训为特色的一家公司。贵公司的先进技术如 13 英寸更纤薄更轻盈的配备有视网膜屏幕的 Mac Book 和 15 英寸显示器 256G 内存的 Mac Book Pro，最新研发的计算机操作系统 Mavericks，完美的多媒体性能以及时尚设计恰恰能满足我们的需求。另外，产品所附带的软件和服务包似乎正是为我们这样的公司开发的。有了这些，我们能及时解决网络连接、密码管理、数据恢复以及计算机更新方面可能遇到的问题。

但是，在看过贵方的报价单，并和其他商家所提供的报价单进行对比之后，我觉得贵方报价偏高。所以我想写信了解一下，如果我们大宗采购的话，贵方能否给予折扣。假如我们一次性购买 100 台，贵方能否提供九折优惠？

另外，我还想咨询贵方的局域网设备情况，因为我们公司很快将要组建一个局域网。贵方能否发给我们相关产品的宣传册以便我们先行了解？

希望尽快得到贵方答复。

你诚挚的

托尼·张

第四单元 产品测试

Text A

测试网页可访问性的五个步骤

测试网页可访问性并不难，只需熟悉 HTML 和 CSS 的相关知识，以及能够预测使用者会遇到的特殊困难。[1] 以下步骤供参考。

第一步：有时无效代码也能让网页正常运行，而且不留任何明显的漏洞，这或许令人惊讶。尽管逻辑上正确，但它的编写方式并不符合标准。然而，有效代码更易于维护，更易于和许多其他技术兼容。[2] 所以，为了减少无效代码可能带来的各种各样的问题，首先一定要确认代码是否有效。

第二步：可访问性自动化测试是整个测试过程中很重要的一步。写文章时，往往要依靠自动拼写检查发现打字错误，即使你还是要亲自浏览并检查确认写的是 Dave 而不是 Cave。[3] 自动化测试能发现人工阅读易于忽视的许多问题。

第三步：不用鼠标而仅用键盘来浏览网页。确保网页上的每个链接、按钮或者其他功能都能通过键盘实现。这一步很有必要，因为并不是所有人都用鼠标浏览网页。

第四步：试着完成一两个网页的功能。如果是一家网上商店，就选一种商品并买下来。如果是一个信息网站，寻找其关键信息。记住——这才是你设计此网页并使用户可访问的原因。

第五步：网页可访问性不仅指能完成一系列请求或者验证它是有效的，网页可访问性还关乎着高质量的设计。因此，最好能请一位屏幕读取和用户界面设计方面的专家测试你的网页，因为他们很熟悉网页的功能、用途以及浏览设计差的网站时会遇到的种种困难，并且能切实解决这些问题。

网页可访问性测试是每个公司和组织设计网页或者计划设计网页时需要考虑的问题。这不仅在道德层面上是正确的，还有利于扩展潜在的客户群，而且这种测试并不难也不昂贵。

Text B

软件测试介绍

软件测试是一种有助于找到正在开发中的软件系统漏洞的过程，以提供给顾客一个无漏洞、安全可靠的软件系统。

软件制造商进行软件测试以确保产品的准确性、完整性和高质量。如果软件运行没有问题，客户将会非常欣慰，从而给制造商带来更多的利润。

软件测试者通过各种输入值，试图找出系统中所有可能存在的漏洞。优秀的软件测试人员技术娴熟，讲究沟通技巧，他们知道如何机智地与软件开发师交流，并说服他们修复所有漏洞。

除了找到软件产品中的漏洞并确认程序符合规范，作为一名测试人员，你还需要编写测试案例、程序、脚本和生成数据；要实施测试程序和脚本，分析测试标准，评估测试结果。除此以外，还需要：

· 在研发早期发现漏洞，从而加速开发进程。

· 减少公司承担法律责任的风险。

· 在产品生产之前发现漏洞和设计缺陷，确保产品成功上市，节约成本、时间以及公司名誉不受损。

软件测试方案描述测试目的、范围、方法和测试的核心任务。准备测试方案的过程是认真考虑如何验证软件产品的可接受性的有效途径。完整的文件描述能帮助不懂软件测试的人了解产品验证的原因和过程。测试方案应该详细、实用，但又不能太过于全面以免测试组以外的人也能读懂。

最后，测试应该按照测试方案中所规定的标准完成。这时根据测试结果来决定所测试软件的成败与否。测试总结报告就要记录这些信息，提供所有测试工作的结果和对这些结果的评估结论。更为重要的一点是，测试总结报告提供了测试和软件质量的全面记录。

第五单元　计算机市场营销

Text A

商业营销：策略开发

营销是企业成功的关键。企业要想在竞争对手中脱颖而出就需要开发一套好的营销策

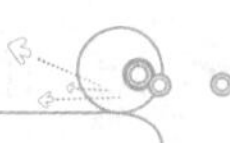

略。[1]仔细的调研、适当的宣传和及时的跟进对建立一个牢固的客户群来说都是十分重要的。[2]

调研市场

你的脑海中是否有一个目标市场？了解你想争取到的客户。调查出你的产品或服务最吸引哪些人群，找到在本区域里哪些公司需要你的产品或服务。认真研究你的目标市场，从而评估那些你所能提供的东西。你可以发放一份简短的调查（不超过10个问题）。这是一个又快又简单的挖掘客户需求的方法。

仔细研究竞争情况能够帮你找到一些你的公司可以填补的需求空白。[3]弄清楚你的公司能提供哪些别的公司无法提供的产品或服务。研究表明有专长的公司比那些试图包揽一切的公司表现得更好。

宣传企业

这一步可以从发放简单的传单开始。任何市场策略都应该包括以下基本要素：你是谁，提供什么，位置在哪里，以及如何联系。

在广告中，尽量通过强调那些只有你的公司才能提供的专长来吸引顾客。焦点放在顾客的一两个特殊需求上，而这一两个特殊需求正是你们能满足的。

在广告中可以包括的其他信息有：开业日期、特价销售、畅销产品、至今已有客户数量、公司口号或标志。

客户跟进

客户跟进是营销中经常被忽视的一个环节。拥有新客户当然好，但是想要真正成功，至少要有一些回头客。而你只需做一些十分简单的事情就可以培养回头生意。

当你接到电话或者收到电子邮件时，务必回复。打一个电话或者回一条消息，感谢他们再次询问。提供优惠券、折扣或者免费礼品（比如有公司标志的钢笔），这些永远都是感谢并鼓励客户再次光临的好方法。

如果照这些步骤去做，你一定会看到你的生意开始成功。一旦你找到了有效的办法，坚持反复使用，这样你很快就会有一个牢固的客户群了。

Text B

淘宝网店主的天堂

你能相信在杭州有一座商务大厦，里面70%的业主都是中国最大的拍卖网——淘宝网的店主吗？这座20层的大厦叫“四季星座”，昵称是淘宝之家，因为有几百个成功的淘宝店的店主租下了楼内的大部分办公室。该大厦坐落在杭州东部最大的服装市场——四季青市场的对面。

据报道，淘宝网生意兴隆，该大厦每天都要寄出10,000多个包裹。他们不但给自己带来了大量的利润，还刺激了当地的经济发展，并提供了大量的就业机会。

“淘宝”的中文意思就是寻找有价值的东西，它提供免费的网上拍卖服务。淘宝之家的大部分经营者都是20几岁的年轻人，主要经营服装、饰品、鞋子等。淘宝网的主要经营模式旨在降低经营者的商业成本，这给生产商和消费者带来双赢局面。消费者常常会买到便宜货或者找到难买的东西。现在，淘宝网主要的消费群体在18至35岁之间。他们追求时尚并拥有相对较高的消费能力。

张楠是淘宝之家的一个网店店主。据说，他以前是在家里经营网上业务的，但经常觉得时间紧迫，因为供给他货源的批发市场离他家太远了。后来，他听说了淘宝之家，那里的大部分租户都是年轻人，而且拿到货源很方便。在核实了位置之后，张楠就决定在那里开一家店。张说，很多店主都毕业于设计院校，他们自行设计产品，然后到杭州郊区的工厂进行生产。这种合作关系有助于降低成本。张说，淘宝之家是一个非常适合大学生创业的好地方，租金便宜，而且允许共用一个店面。

第六单元　计算机物流

Text A

包装的作用

包装是为配送准备货物的技术。近年来，包装的重要性逐渐被认可，今天，广泛使用的包装已成为产品市场竞争的主要竞争点。包装与一体化物流的关系最明显体现在运输上。

包装可分成工业性包装和消费性包装两大类。一般来说，消费性包装主要为了盛装商品，促进商品销售，方便顾客使用，而对物流运作价值不大。但是工业性包装却对物流成本与物流生产力有很大的影响。为符合综合物流的要求，工业性包装需要具有以下功能。

第一，它必须防止商品在搬运、存储和运输过程中受到损坏。包装品在运输过程中，随时会招致损坏，损坏原因有震动、撞击、刺破或挤压。因此，包装设计和采用的材料必须结合起来以达到防范要求，而又不会因过多防范措施导致成本增加。有时所设计的包装品具有合适的材料承载内容但是却不能达到必要的防范效果。要满意地解决包装问题，就需要根据预期的总体状况，判断货物所允许发生的货损程度（因为在大多数情况下，不会为了完全避免货损而使成本过高），然后找出一种包装，供设计和选材符合这些要求。

第二，包装应能促进物流效率。包装不仅影响生产和销售，而且也影响综合物流活动。例如，包装尺寸、形状和材料的种类和数量以及商品在仓库里堆放的方式。同样，包装的大小和形状也影响到产品装货、卸货和运输。产品搬运越容易，运输费用就会越低。因此，如果包装是为了物流作业的高效而设计，那么整个物流系统的运作都会受益。

物流包装的第三个重要的功能就是通信或信息传播。包装最显著的信息传播作用就是收货入库时查验包内货物、分拣和目标货物的查验。典型的包装信息包括生产商、产品、容器型号、数量以及通用商品代码（UPC）。包装追踪便易也很重要。内部动作的有效性和客户数量的增长均要求在产品通过物流渠道时能够被跟踪。射频识别系统的广泛应用，使这一切得以实现。该系统在包装品、容器或运输工具中内置一块计算机芯片，便可以使

其在经过配货站和运输出入口的各检查点时，包装容器和包装内容能够被扫描并得到查证确认。物流包装的最后一种信息功能就是在如何搬运货物和如何防止可能的损坏方面提供指示说明。例如，如果产品具有潜在的危险，像鞭炮，乒乓球之类，其包装材料或辅助材料必须提供相关说明以避免受潮、受震、受热等危险情况发生。

除去商品保护的功能，包装还应该易于处理，方便存放，容易识别，安全，且拥有把空间最大化利用的外形。由于包装的存在，使得产品能在任何时候和任何地方都可得到，这给消费者极大的自由选择。

Text B

影响货物包装类型的因素

包装在具体物理环境中的功能受湿气、温度极限、机械撞击和震动的影响。无论环境状况如何，包装都要能够保护产品，使之保持用户可以使用的状态，直到产品到达最终消费者手中。在确定包装类型的时候，有若干因素要考虑。

(1) 物品价值。高价值货品通常比低价值货品有更宽泛的包装形式。如果包装不合适，在涉及承运人的责任、收货和投保适当的货物运输险种时，可能会遇到麻烦。此外，投保油画等高价值货物就需要充分的安全措施，同时也会产生更多的运费。

(2) 运输属性。即运输的类型与长度。我们必须考虑到装运方式，也许是拼箱或整箱。装卸搬运频率越高的货物要求包装得越结实。倒装程度大的货物也要求更结实的包装。有些运输方式，尤其是空运和国际标准化货柜运输通常要求相对严格的包装。这是他们服务的一个突出的营销特征。空运在运输途中尤其鼓励将货品捆扎在托盘上进行运输。

(3) 货品属性。这关系到所运输货品的特征和它们遭受各种损失或破坏的可能性。须知包装同时要起到防盗和防破坏的保护功能。货品和包装要求同时决定了货品的包装类型。散装货物需要很少或不需要包装。但是大多数货物都需要充分的包装。

(4) 运输过程中温度的多变性。在运输过程中温度的变化可能相当大，要考虑到物品呼吸、避免货品过度浓缩或挥发等因素。如果必要还要向航空公司或船方征求意见。

所以，包装并不仅仅是为在运输中保护产品和降低商品损坏而设计的，它还用于防止盗窃和帮助营销。当然，不仅仅看是否选择了恰当的包装类型，还要看是否运用了合适质量和形式的集装箱。

第七单元　计算机网络

Text A

现代企业网络营销策略

随着互联网的日益发展，如何利用这个跨地域的平台来打破资源障碍、优化资源组合、提升自身品牌价值，已经成为很多企业关注和研究的问题。

产品策略

企业可以通过分析网上的消费者们的总体特征来确定最适合在网上销售的产品。要明确企业产品在网络上销售的费用要远远低于像计算机软件等一些产品的其他渠道的销售费用。在网上销售要比通过其他渠道方便得多，自然花费也就低得多，这样无形中降低了企业成本，提高了企业产品在市场上的竞争力。

企业应利用网络上与顾客直接交流的机会为顾客提供定制化产品和服务。[1]同时企业应及时了解消费者对企业产品的评价，以便改进和加快新产品的研究与开发。另外企业在开展网络营销的同时，可以降低创新风险，减少开发费用。

价格策略

价格是网络营销中最为复杂和困难的问题之一，因为价格对于企业、消费者乃至中间商来说都是最为敏感的话题。网上销售可以使得单个消费者可以同时得到某种产品的多个甚至全部厂家的价格，以做出购买决策，这就决定了网上销售的价格弹性较大。因此，企业在制定网上销售价格时，应充分考虑检查各个环节的价格构成，以期制定出最合理的价格。

因为网上价格随时会受到同行业竞争的冲击，所以企业可以开发一个自动调价系统，根据季节变动、市场供需情况、竞争产品、价格变动、促销活动等因素，在计算最大盈利基础上对实际价格进行调整。[2]同时还可以开展市场调查，以获得有关信息来对价格进行调整。

促销策略

网络广告是目前较为普遍的促销方式。网络广告不像其他传统广告那样大面积地播送（“推”），而是让消费者自己选择（“拉”）。网络的强大功能几乎囊括了所有媒体广告的优势。企业在做广告策划时，应充分发挥网络的多媒体声光功能、三维动画等特性，诱导消费者尽可能做出购买决策，并实现开发潜在市场的目标。[3]

利用网络聊天的功能开展消费者联谊活动或在线产品展销活动和推广活动。这是一种调动消费者情感因素，促进情感消费的方式。在这方面成功的典型是在线书店亚马逊，它在网站下开设聊天区以吸引读者，使其年销售额达到 34%的递增，其中有 44%是回头客。早在 1996 年其销售额就突破了 1700 万美元，充分展示了网上促销的魅力。

通过线上资料库联网，与非竞争性的厂商进行线上促销联盟，增加与潜在消费者接触的机会，这样，一方面不会使本企业产品受到冲击，另一方面又拓宽了产品的消费层面。将网络文化与产品广告相融合，借助网络文化的特点来吸引消费者。

渠道策略

结合相关产业的公司，共同在网络上设点销售系列产品。采用这种方式可增加消费者的上网意愿和消费动机，同时也为消费者提供了便利，增加了渠道吸引力。[4]

在企业网站上设立虚拟店铺，通过三维多媒体设计，形成网上优良的购物环境，并可进行各种新奇的、个性化的，随一定时期、季节、促销活动、消费者类型变化而变化的店

面布置，以吸引更多的消费者进入虚拟商店购物。

网上交易还有很多问题没有解决，如网络付款的安全性，但随着网络技术的日新月异，网上交易必将会越来越完善。

Text B

成为顶级销售员

“在电话推销中，问错了问题可能会立即毁掉你开发新业务的机会，而正确开场问题将会给你带来很多机遇。”英国招聘教练罗伊·里佩尔说。

作为招聘主管，里佩尔认为要好好利用推销电话，这至关重要，在电话中也可以问一些开放式的问题，例如：

- 你能告诉我一点关于你在 X 公司的工作性质吗？
- 你们过去的招募怎么样？
- 为什么有人想去你们公司工作？

不要说：

- 你们现在招聘吗？
- 你们有其他招聘计划吗？
- 在你的公司有很好的晋升潜力吗？
- 有培训课程吗？
- 你打算在不久的将来再招聘吗？

备用市场的应聘者

在杰西的招聘培训录像中，里佩尔为备用市场的热门候选者提供了一个脚本。

他说，打电话时你不能说：“嗨，今天是你的幸运日。我这里有一个拥有四年销售经验的候选者在找工作。”

根据里佩尔所说，引入一个应聘电话营销的最佳方式，是确保应聘对象能为招聘单位做些什么。

他建议这样说：“我的应聘者在过去四年一直都是他们公司的顶尖销售人员，他在这一时期获得了两次升职机会，他是唯一快速被邀请到公司管理层的销售人员。”

“这位应聘者的推荐信是我见过的最好的，他表示有兴趣去你们公司您所在的部门工作。所以我今天下午联系了您和另外三家公司谈关于这位应聘者的事。他下周已经有两个面试计划，所以……我什么时候能拟订出您的面试时间？”

“调整脚本来适合你的个人风格和应聘者的具体情况是很重要的，因为业务发展要求中‘照本宣科’是人力资源决策的五大忌之一”，里佩尔说。

根据招聘主管杰西的研究，其他大忌包括：

- 过分热情。
- 电话过于频繁。
- 带有攻击性的问题。

• 没有原因的打电话。

业务发展建议

根据里佩尔：

• 每一个你打的电话都应该使你销售策略的步骤前进一步。

• 找到一天中的最佳时间和要打电话的相关场所，然后在应聘电话中尽情发挥。

• 在考虑准备电话和打电话前要首先建立你的自信。

• 你每天打的第一个电话要是一个温暖的电话。

• 在每天结束的时候花至少 30～40 分钟来计划下一步的行动。

第八单元　电子商务

Text A

什么是电子商务?

电子商务是指企业利用互联网与信息技术，以提高经营效率为目标，在企业经营全过程的各个环节实现自动化、电子化、无纸化，这些环节包括市场研究、谈判、配送、生产、资金支付和售后服务等。[1]

电子商务本身不是一种纯技术，而是互联网与信息技术在商务中的运用过程，以纸介质为基础的用于信息交流、传递、处理及存储的工具被以电子媒体为基础的工具所取代。[2]

在传统经营环境中，一个公司要花相当多的时间和精力来调查和决定生产与销售什么或者生产与销售多少的问题，然后，在媒介上（如电视、广播、路边广告牌等）做广告，与买家通过电话、电传、邮件或直接见面方式接触；看过货后，再讨价还价，然后买方下订单或与卖方签订协议，卖家准备好货物后将货物送给买方，买方以现金或支票付款。在以上过程中，所有信息是通过电话、邮件、传真或电报方式，或是以近十几年来使用的 EDI 系统进行交流的。EDI 是一种计算机网络系统，可处理一些特定的数据（如支票、合同、信用证等）在特定的合作伙伴之间的交换，建立一个 EDI 系统花费甚巨，而且使用起来相当复杂。所以，正如你所看见的一样，在传统经营环境里，人们需要花费更多时间和精力进行信息交流，有时还会出现错误和延误。

在电子商务环境中，人们利用万维网或互联网做生意，买卖双方的接触不再需要面对面，而是通过网络。在网上，买卖双方都有更多机会了解对方，不仅是一对一，而且可以多对多。无地域限制，无时间限制，双方都能找到称心如意的对家。买家利用浏览器根据想去的网站的域名到网站或电子商店的首页上寻找产品目录，而且可以在那些网页上看到产品的包装、外形、款式、色泽和价格，有时，甚至还可在网上议价。如果满意这些货物，可以点击选择键或以电子邮件的方式下订单。卖方立即就可收到信息，如果他们接受订单，他们就用万维网或互联网去安排货物生产，配送和货款支付（称为电子资金转移），所有数据被及时传递以使企业经营变得更有效率。

Text B

EBay 贸易

1995 年 9 月，Pierre Omidyar 创建了在线拍卖企业即著名的 eBay 贸易。他的目的是创建一个为个体或由个体进行商品和服务销售的虚拟市场。

为了使业务开展起来，Omidyar 和他的共同创始人 Jeff Skoll 得到了哈佛商学院优秀研究生梅格·怀特曼的帮助。梅格·怀特曼成立了由工业界领袖如 Pepsico 和 Disney 为高级执行人员的管理团队，为 eBay 创建了坚定的使命，那就是把商界的人们联系起来——并不是向他们出售产品。更为合理的说法是：他们创建了基于互联网的人对人的市场，在这个市场环境下卖方可以列出待出售的商品以吸引买方进行竞标。

eBay 贸易刚起步，网站立即成为一个收藏品拍卖的畅销渠道，但不久它就涉足到其他市场领域，如汽车、工商业设备和消费电子设备等平均销售价格较高的商品。商品平均销售价格的上升是 eBay 增加销售收入的一个重要部分，因为 eBay 是按照销售收入的百分比收取交易费的。产品平均销售价格越高，eBay 促成的每百万交易量所获得的利润越多。

eBay 开创了自动在线的人与人的拍卖贸易，并使之国际化。以前，这种贸易是通过库房销售、收集物展览、跳蚤市场和分类广告的形式进行的。在线市场使得买方容易浏览商品信息并且能使卖方注册后把自己的销售信息发布在 eBay 上。

浏览和竞标拍卖是免费的，但是卖方通过 eBay 销售商品是要付交易费的。交易费有两类。一是注册费或登记费。当卖方在 eBay 上发布商品信息时，eBay 根据卖方发布的该商品的开拍价收取不退还的注册费。二是最终交易费。拍卖一旦结束，eBay 就收取最终交易费。最终交易费通常根据商品的最终销售价格的 1.25%～5%收取。

eBay 也可以通过提升拍卖的特征来提高发布费用，包括采用高亮显示或粗体标题，商品的特征和其他的使卖方能增加其商品可视度的方式。一旦交易结束，eBay 就以电子邮件的形式通知买卖双方。买卖的成交取决于卖方和买方，eBay 最后收取不包含货款和运费的最终交易费。

第九单元　计算机安全

Text A

电子商务的安全

在电子商务中一些令人担忧的事情

在网络给人们无尽方便与机会的同时，它也给人们带来了风险。由于电子商务是在因特网上进行的，而且它是如此开放的系统，所以，当所有资料信息必须以某种方式电子化被处理、传递和储存时，会发生一些风险。与网络和电子商务应用有关的风险可以以多种形式发生，如数据被偷窃、被讹用、被滥用、被改动或被错误处理，计算机系统和网站也可能因为被攻击而导致处理系统无法正常工作，硬件或软件被非法使用，所以这些破坏可使企业或用户受到严重伤害或经济损失。

为何有这么多不愉快的事情发生呢？这有两个原因。一是网络本身就是新事物，它发展非常快，前景诱人，多功能服务和在网上设立的爆炸式的增长速度吸引越来越多的公司和个人像淘金热般地加入进来。但在许多方面，如信息技术、网络协议、语言、网络标准、制度与法规等网络运作有关的方面远远没有完善。[1]

另一个原因是人们如此急切地投入网络和电子商务应用，以至于他们来不及考虑可能涉及的细节。所以他们根本不可能随时做好准备以给网上犯罪者迎头痛击，因为他们没有足够的知识与经验应对可能面对的那些风险，不知如何发现那些犯罪者的企图。[2]于是也没有适当的控制手段对付他们。

网上犯罪的沉痛教训将人们的注意力转向改善网络及电子商务安全状况，因为他们期望在一个安全环境下处理他们的电子商务。[3]总地来讲，安全电子商务的基本要求应能概括为如下内容：

第一，保护电子交易的隐私。这意味着所有电子资料应避免非授权暴露。交易资料有许多种，如广告、产品目录、客户的个人资料或机密数据。所以，我们应采取不同的方法去保护不同种类的资料。隐私的另一个含义是，电子交易和交易内容应该并且不能被任何人知道，除了贸易伙伴之外。目前，用于确保隐私的安全技术是加密、防火墙和口令。

第二，保护电子交易的内容完整。这意味着保护和证明在电子交易中记录下来的资料的项目与内容确实是双方认同的内容，并且，应在处理与保存过程中保持资料的完整，不在任何非授权方式下改动它们。

第三，安全确认系统。当合作伙伴送来的电子信息被用户接收时，用户和发送者的身份都需要证明以确认接、收双方是否是他/她所自称的那个人。[4]通常，他们的身份可通过验证他们的口令、数字签名或由授权的第三方数字证书来证实。另一件需要证实的事是这笔交易的存在，这称为“不可拒绝或否认”。

总地来说，电子商务安全的要求是：电子交易的内容和隐私必须被保护，以防止在电子数据的处理、传送、储存过程中被截获、滥用、改动、删除或歪曲。

Text B

计算机网络攻击和防范

网络安全是计算机信息系统安全的一个重要方面。如同打开了的潘多拉魔盒，计算机系统的互联，在大大扩展信息资源的共享空间的同时，也将其本身暴露在更多恶意攻击之下。如何保证网络信息存储、处理的安全和信息传输的安全的问题，就是我们所谓的计算机网络安全。信息安全包括操作系统安全、数据库安全、网络安全、病毒防护、访问控制、加密和鉴别七个方面。

下面是网络安全的解决方案。

1. **入侵检测系统部署**

入侵检测能力是衡量一个防御体系是否完整有效的重要因素。强大完整的入侵检测体系可以弥补防火墙相对静态防御的不足，对来自外部网和校园网内部的各种行为进行实时检测，及时发现各种可能的攻击企图，并采取相应的措施。

2. **漏洞扫描系统**

采用目前最先进的漏洞扫描系统定期对工作站、服务器、交换机等进行安全检查，并根据检查结果向系统管理员提供详细可靠的安全性分析报告，为提高网络安全整体水平产生重要依据。

3. **网络版杀毒产品部署**

在该网络防病毒方案中，我们最终要达到一个目的就是：要在整个局域网内杜绝病毒的感染、传播和发作。为了实现这一点，我们应该在整个网络内可能感染和传播病毒的地方采取相应的防病毒手段。同时为了有效、快捷地实施和管理整个网络的防病毒体系，应能实现远程安装、智能升级、远程报警、集中管理、分布查杀等多种功能。

4. **网络主机操作系统的安全和物理安全的措施**

网络防火墙作为第一道防线，并不能完全保护内部网络，必须结合其他措施，以改善该系统的安全水平。根据级别从低到高，即主机系统物理安全、核心业务系统安全、系统安全、应用服务的安全和档案系统的安全。同时，主机的安全检查和错误修复，以及备份安全系统作为补充的安全措施，这些构成了整个网络系统。第二道防线防止防火墙被突破以及来自内部的攻击。系统备份是最后一道防线网络系统，用于攻击后系统还原。防火墙和主机安全的措施是整个系统的安全审计，入侵检测和响应处理器构成整体安全检查和应对措施。从网络系统的防火墙、网络主机，甚至直接从网络链路层的提取网络状态信息，输入到入侵检测子系统。入侵检测系统按照一定的规则，以确定是否有任何入侵的事件。如果发生入侵，它会紧急处理措施，并产生一条警告消息。此外，该系统的安全审计还可以作为未来后果的攻击行为，并可以处理系统的安全政策以改进信息来源。

总地来说，网络安全是一个综合性问题。它涉及技术、管理、使用和许多其他方面，包括它自己的信息系统安全问题，有物理和逻辑的技术措施。作为一种技术，它只能解决问题的一方面，而不是万能的。

第十单元　计算机故障及排除

Text A

计算机软、硬件的一般故障

硬件故障是由计算机硬件引起的，涉及计算机的主机系统、存储器、键盘、显示器和显示部件、磁盘驱动器和控制部件、电源与供电部件等。

软件故障，狭义上包括误操作、驱动程序冲突、初始化设置错误、软件自身存在 Bug 等因素。广义上，软件故障涵盖了以计算机为核心、以网络为外延的现代化商务办公体系：系统及应用程序安装（含新机软件安装）、办公自动化及电子邮件处理、病毒及木马、硬件及网络、网络驱动及程序、应用软件更新、打印机等的故障。[1]

硬件故障和软件故障的原因

硬件本身质量不佳。使用劣质的硬件替代系统原有的硬件。不合规格的硬件与原有的系统相结合，非常容易引起系统的紊乱甚至是内部线路的短路和接触不良。

人为因素影响。在实际的操作中，使用者的操作过程不符合系统的运行，导致硬件出现故障。

适用环境影响。任何一个环境因素对于机器的影响都是非常巨大的，计算机运行环境超过了硬件允许的极限值都会严重影响计算机的性能，造成硬件故障。[2]

软件故障比较频繁发生，一般由病毒感染、安全漏洞、木马黑客、非法操作、异常关机、流氓软件滋扰、文件损毁、授权加密、硬件变化等几种情况引发。

Text B

预防电脑病毒

你肯定生过病。生病通常是因为感染了病毒。握手、打喷嚏都能让病毒繁殖传播。计算机病毒亦然。计算机病毒不是微生物而是计算机程序。黑客设计的病毒程序会像疾病一样在计算机间传播。一旦计算机感染了病毒，它会删除文件、私发邮件，甚至把私人信息泄露给犯罪分子。计算机病毒可以细分为三种：普通病毒、蠕虫和木马。

先来说说普通病毒。它通过计算机之间的文件共享来传播。通常会在附件或 U 盘中出现，一打开这种文件，计算机就会中毒。然后复制，等着感染下一台计算机。和人生病一样，有时候文件是否感染病毒也很难“确诊”。所以，最好的防护措施就是安装杀毒软件，能防止计算机中毒，也能在发现病毒时及时清除。

蠕虫则更可怕，即使我们什么也没操作，它也能传播。只要计算机联网，蠕虫就会感染网上的病毒。不管是局域网，还是在整个互联网中。蠕虫擅长走“后门”。就是通过计算机软件的漏洞入侵。一旦被感染，蠕虫还会寻找其他计算机的“后门”，一路披荆斩棘地破坏下去。最好的防护措施是更新系统。门关好了，蠕虫就爬不进来了。

木马则是最阴险狡诈的病毒。像传说中的特洛伊木马，它也善用诡计，一不小心就可能从网上下载木马。木马病毒会伪装成游戏或者常用软件，有些木马还会给计算机开新的“后门”，让“犯罪分子”轻易侵入你的计算机获取信息。要防木马，你必须只在信任的网站上下载软件，提高警惕，时刻注意网络安全，就像咳嗽要掩口，饭前、便后要洗手一样。还要经常更新计算机软件，安装杀毒工具。一边防患于未然，一边“亡羊补牢”。另外，陌生的链接、附件、文件也不要碰。除非能确认它们的合法性。

第十一单元 售后服务

Text A

惠普[1]退货和换货

退换货政策

退换期限

我们接受交付后最长 21 日内的退换请求。超过此 21 天的期限后，您可从惠普客户服务部获取产品售后支持。[2]

信用额度

退换货的信用额度将在我们确认您购买产品的收据后立刻进行处理。请知悉您的银行可能需要 5～7 个工作日对您的账户进行处理。

退款

我们承诺含税的全额退款。首次运费以及处理变更费用不予退还。当您订单中的部分商品被退货时，对整单有效的折扣将不再适用（如礼品、大宗购买）。在这种情况下，折扣部分将从您购买产品支付的金额中扣除。为了得到全额退款，购买时的全部东西都需要退还。

惠普客户服务为您提供不间断的全面服务与支持，请您：

- 访问“支持和驱动”以获取在线支持、产品信息、软件以及驱动。
- 通过以下电话联系我们：

惠普产品：（800）4746836

康柏产品：（800）6526672

退换程序

进行产品的退换前，您需要首先通过拨打（800）9994747 联系我们，获取一个退换货品授权号码（RMA）。

1）您会收到一封含有联邦快递运输标签的电子邮件。

2）5 天之内请打印这个标签。

3）将要退还的产品和所有包装箱中的附件放回原包装中。

4）用新标签替换原有的运输标签。

5）用粗体字在包装上注明“RMA（退换货品授权号码）”。

6）用联邦快递将货品发出。

注意：

1）我们承担退货运输费用。

2）您可以在订单状态中查看您退换货的进度。

3）对于换货，我们需要在收到您寄回的货品后才能将新货品发出。

4）不遵守此流程可能延误您的信用额度或者换货进度，同时您需承担导致损失的风险以及寄回产品所需的费用。

Text B

如何投诉？

当人们需要抱怨产品或服务不善时，一些人选择书面投诉而不是当面抱怨。你会采用哪种方式？请给出具体原因及实例来支持你的答案。

当我想要做一个关于抱怨缺陷产品或服务欠佳的投诉时，我选择书面投诉。书面投诉让我用合乎逻辑的方式组织我的观点。如果我很不满意我的遭遇，我会清楚地写出原因。我不希望我的抱怨引起任何混乱。我会写出我的不满并且给出实例。这是使每个人都明确我在抱怨何事的最好办法。

书面投诉也能确保不显得太情绪化。如果你觉得你受到不好的待遇或被欺骗，很容易情绪失控。大发脾气，肯定让你在投诉中败下阵了。在当时大发脾气会令你很发泄，但最后它令被你训斥的人更厉害地对待你。这种处理方式不会令他们同意为你提供帮助。

你要处理的另一个问题是：如果你当面投诉，你必须跟在那里的人当面对话。他或她可能不是对缺陷产品或服务不善应该负责的人。除非是一个非常小的业务处理，通常接待投诉的人都不是负责人。对这些人发脾气是不公平的，同时也不会得到满意的处理结果。你要当着负责人的面抱怨。所以书面投诉是解决问题的最好办法。

书面投诉并以挂号形式邮信是很好的书面证明。这明显表示你试图在一定时间内以合理的解决方式解决事情。这样，如果你需要采取进一步行动，你就有物证。

书面投诉有组织性、高效性和公平性的优点。这就是为什么我选择书面投诉而不是当面抱怨。

第十二单元　计算机的前景

Text A

21 世纪人与计算机的融合

2020 年的教育：向计算机辅助的自学发展

学生们通常都有自己的计算机：那是一种便笺一般薄的装置，不到 1 磅重，以及适于阅读的高清晰度显示器。学习材料可利用无线通信获得。由教师本人指导一群学生的传统教学方式依然盛行，但学校的发展越来越依靠软件教学。教师们主要致力于学生学习兴趣的激发、心理健康和适应社会等问题。

2020 年的商业：虚拟商业[1]得到发展

至少有一半交易在网上进行。工作群体的地域分隔趋势日益明显。在不同的地方生活和工作的人们可以成功地协作。平均每家拥有两台以上计算机，其中大多数设置在家电或通信系统中。家用机器人已经出现，但是尚未被完全接受。

2020 年的保健：电子护理人员

远程医疗[2]被广泛使用：医师们利用远程视觉、听觉和触觉检查为患者进行诊断。以计算机为基础的模式识别广泛应用于解释成像数据和其他诊断程序。患者的终生病例都保存在计算机数据库中。

2020 年的艺术：媒体加强了表现力

计算机屏幕是视觉艺术的选择媒介，它正逐渐成为艺术家与他们的智能艺术软件的工作的结合点。技术能使非音乐家创造出音乐，如利用人脑电波作曲的声控作曲系统和软件。作家则使用声控文字处理系统和文体改进软件。

Text B

视频网络

几年前，只有几家网站发布用户上传的视频。如今这类网站已经超过 225 家，形成了一个发布业余爱好者和专业人士的视频作品的平台。

接下来会有什么新的发展呢？视频内容的发布曾经是由一些广播网络和有线电视公司所控制的。但是当视频内容与分散的、以用户为中心的万维网相逢时，我们是否看到了“视频网络”诞生的曙光？这种新媒体将重新定义媒体的概念。专家预测，几乎每个面向消费者的机构——从电视台和体育团队到软饮料和洗洁剂制造商——都会迅速在网上开发视频呈现。而且还不仅如此。如果视频发布业务以近几年网站和博客的发展速度增长，那么它对传统的广播公司、企业和用户意味着什么？

业余爱好者的作品目前占据主导地位。大多数网站提供免费上传，并试图通过广告或分销传统广播网络、制片商和其他内容合作商的商业视频内容来盈利。

由于业余摄影师通常会避开主流的内容和制作风格，寻找一些自己感兴趣的内容，因此用户制作的内容从发人深省的访谈到喜剧模仿秀到街头斗殴的残酷场面，应有尽有。其中最流行的视频类型是体育视频（比如滑雪或赛车）、模仿秀和讽刺短剧、评论、访谈、电视短剧和纪录短片。

由于视频爱好者的市场庞大，关键问题是：“如何建立起相关业务？如何为这类媒体和观众创造产品？”

媒体公司正一马当先。每个传统的新闻机构都在开拓网上视频业务，报纸也和电视广播商一样跃跃欲试。市场数据表明由于因特网作为一种新闻来源的兴起，传统媒体的受众日渐减少。忽视这一数据的代价没有人能承受得起。最近的一项研究表明，美国每天有 5000 万以上的人把因特网作为主要的新闻来源。

专业的广播公司也不甘示弱；前因特网创业家，如今在南加州大学的安娜伯格传媒学院就职的乔纳森·塔普林预测，事实上，大多数专业有线电视台的盛世即将结束。“今天我们有大约 440 家小的有线电视台；在 5 年之内，其中 300 家将消失，剩下的就转向网络业务。例如，探索频道正在因特网上推出新的专业频道，因为它没有足够的资金在其他媒体推出新频道。”

Appendix B

练习答案

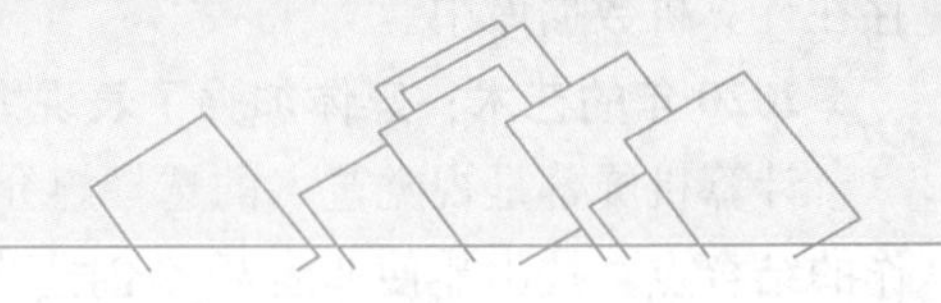

Unit 1

Text A

Ⅰ. **Answer the following questions briefly according to the text.**

1. In 1946.
2. They were the first generation of computers, huge, heavy, expensive and slow, as well as using much more power than today's, but they still made great contributions to computer science, such as the concepts of stored programs, random access.
3. The invention of transistors.
4. Integrated circuits meant that huge complicated circuits and millions of their elements were only made on a small semiconductor chip.
5. The unprecedented access to information provided by computers has changed society's privacy landscape. Credit cards, banks and telephone companies record users' business activities. Internet firms that provide free services such as online searches, maps, and e-mail save information typed in by users and sort it by unique identification numbers in the machines.

Ⅱ. **Fill in the table below by giving the corresponding Chinese or English equivalents.**

electronic digital computer	数字电子计算机
vacuum tubes	电子管
computer science	计算机科学
transistor	晶体管
integrated circuits	集成电路
semiconductor chip	半导体芯片
middle class	中等收入家庭
online searches	在线搜索
Internet firms	网络公司
biological computer	生物计算机

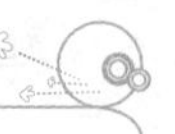

Ⅲ. **Read the text again and fill in the blanks with the information you have got in the text.**

1. the first generation；made great contributions to
2. huge complicated circuits；millions of their elements
3. unprecedented access to
4. science and technology；emerge out

Ⅳ. **Choose the best one from the items given below to complete the following passage.**

1. B　　2. C　　3. A　　4. D　　5. C

Text B

Ⅰ. **Read the passage above and decide whether the following statements are true (T) or false (F).**

1. F　　2. T　　3. F　　4. F　　5. F

Ⅱ. **Match the words or phrases on the left with their meanings on the right.**

1. D　　2. H　　3. A　　4. F　　5. J
6. I　　7. G　　8. B　　9. E　　10. C

Ⅲ. **The following is a list of terms of computer. After reading it, you are required to find the items equivalent to those given in Chinese in the table below.**

A—9；　B—7；　C—8；　D—5；　E—6；
F—1；　G—2；　H—3；　I—10；　J—4.

Ⅳ. **Translate the following paragraph into Chinese.**

输入设备：它把人们能够理解的数据和程序翻译成计算机能够处理的形式。最常见的微机输入设备是键盘和鼠标。计算机上的键盘看起来像一个打字机键盘，但它含有一些特殊的键。鼠标是一个可以在桌面上滚动的设备，它给出屏幕上插入点或光标的位置。一个鼠标有一个或多个用来选择命令的按钮，鼠标也可用来画图。

Writing

1. Jameson　　2. the 5th Avenue，New York City
3. 1902－32323　　4. (44) 1902－32123
5. hjameson@google. com

Unit 2

Text A

Ⅰ. **Answer the following questions briefly according to the text.**

1. After Steve Jobs return to Apple.
2. Characterized by its curvy，translucent backing that differentiated it from every computer in its class.

3. The iPod line which designed by Ive shocked life back into Apple and made it an industry leader in multimedia ventures.
4. The user define the device rather than the device define the user.
5. Undoubtedly magical.

Ⅱ. Fill in the table below by giving the corresponding Chinese or English equivalents.

design patent	设计专利
Post-PC	后个人计算机时代
Newton device	牛顿机
printer tray	打印机纸盘
flat panel computer	平板电脑
designer's prototype	设计模型
cathode-ray-tube computer	阴极射线管计算机
media player	多媒体播放器
unique approach	独特的理念
design principle	设计原则

Ⅲ. Read the text again and fill in the blanks with the information you have got in the text.

1. Concealed from；next-generation electronics；tucked away from
2. utilize his talents
3. met Jobs' expectations；causing his pain
4. played an important role in；mindbogglingly simple
5. crowning achievements；revealed her adoration

Ⅳ. Translate the following paragraph into Chinese.

那个企业家拥有一个超过 2000 名员工的大公司。他经常使用谷歌来为他的国际贸易寻找信息。他的秘书常常帮助他做处理信函、安排各种会议等的事情。他经常参加各种各样的早餐会晤，这些早餐会晤很少持续一个小时。在他看来，餐桌礼仪十分重要。他要求公司的设计师们尽最大努力让产品具有吸引力。

Text B

Ⅰ. Read the passage above and decide whether the following statements are true (T) or false (F).

1. F　　2. F　　3. T　　4. F　　5. T

Ⅱ. Complete the following sentences by translating the Chinese in the brackets.

1. punching a button and doing everything from start to finish
2. make the whole system inflexible
3. the level of programs available to the user
4. the Access database breaks
5. A wise selection of the tools that are available

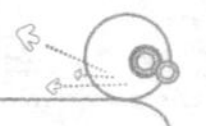

Ⅲ. Choose the best one from the items given below to complete the following passage.

1. B 2. A 3. D 4. C 5. A

Ⅳ. Translate the following paragraph into Chinese.

运行在计算机上的最重要的程序是操作系统。每台通用计算机必须有一个操作系统来管理其他程序。操作系统完成一些基本任务，例如从键盘上辨别出输入数据，向显示器发送输出信息，跟踪磁盘上的文件和地址目录，控制诸如磁盘驱动器和打印机等外围设备。

Writing

MEMORANDUM

To：1) Quality Control Manager

From：Luo Rui

Date：2) 5 July 2008

Subject：3) About the New Machine

At the recent test，I discovered that 4) there were some problems with the new machine（新机器有问题）. May I suggest that the production of the new machine be stopped to ensure 5) the quality of the product（产品质量）?

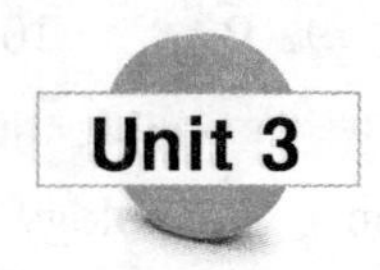

Unit 3

Text A

Ⅰ. Each IT companies has several departments. Match each department with its main responsibility.

A—7 B—4 C—2 D—6 E—8 F—1 G—3 H—5

Ⅱ. Answer the following questions briefly according to the text.

1. Apple Inc. is refreshing its iPad lineup and slashing the price of its Mac computers ahead of the holiday shopping season.
2. The iPad Air weighs just 1 pound，compared with 1. 4 pounds for the previous version. And it's tablet is called "screaming fast iPad."
3. It has a retina display designed to give it a clearer，sharper picture and the same 64-bit chip that powers the iPad Air.
4. 9 hours.
5. Yes，it's free.

Ⅲ. **Read the text again and fill in the blanks with the information you have got in the text.**

1. along with a slew of
2. it is eight times faster than
3. the previous version
4. is available

Ⅳ. **Translate the following paragraph into Chinese.**

到目前为止，世界上还没有哪个计算机行业或者其他任何行业的领袖能够像乔布斯那样举办一场万众瞩目的盛会。在每次苹果推出新产品之时，乔布斯总是会独自站在黑色的舞台上，向充满敬仰之情的观众展示又一款“充满魔力”而又“不可思议”的创新电子产品来，他的发布方式充满了表演的天赋。计算机所做的无非是计算，但是经过他的解释和展示，高速的计算就“仿佛拥有了无限的魔力”。乔布斯终其一生都在将他的魔力包装到设计精美、使用简便的产品当中去。

Text B

Ⅰ. **Read the letter again and choose the best answer to each question.**

1. D　2. C　3. D　4. A

Ⅱ. **Match the following terms with their Chinese meanings.**

1. E　2. C　3. D　4. A　5. H
6. J　7. F　8. G　9. B　10. I

Ⅲ. **Complete the following sentences by translating the parts given in Chinese.**

1. With the development of science technology
2. share information and resources through network communication
3. complete automation
4. arithmetic and logic calculation
5. a digital encryption and signature device

Writing

Employment organization	Avon (Guangzhou) Company Limited
Card holder	Liu Dong
Title/position	Purchasing Assistant
Address	422 Huangshi Road, Baiyun District, Guangzhou
Postal code	510426
Fax	020 - 86625598
Telephone	020 - 86453599
E-mail	ld@sina. com. cn

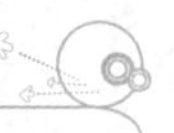

Unit 4

Text A

Ⅰ. **Answer the following questions briefly according to the text.**

1. The familiarity with HTML and CSS and the ability to realize the special difficulties faced by users.
2. By validating the code.
3. It can help finding many problems which could easily be missed by reading the code.
4. Because not all people use the mouse to browse websites.
5. It can increase the potential customer base.

Ⅱ. **Fill in the blanks with words or phrases that match the meanings in the right column.**

1. familiarity	2. invalid	3. intimate
4. eliminate	5. accessibility	6. browse

Ⅲ. **Read the passage again and match each step with a proper title.**

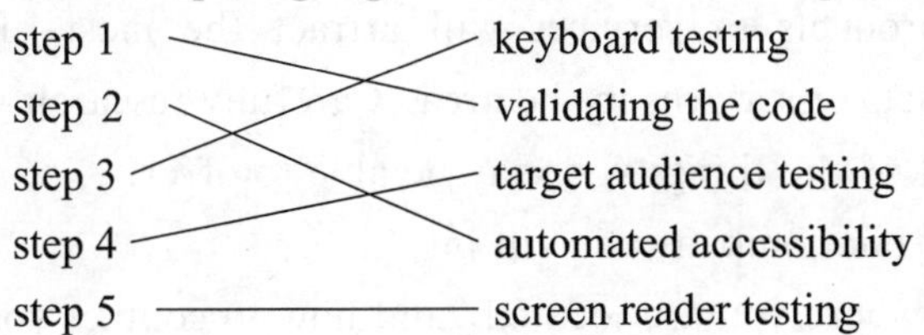

Ⅳ. **Fill in each blank with the appropriate form of the word given in the brackets.**

1. network	2. accessible	3. invalid
4. navigation	5. intimately	6. was frustrated

Ⅴ. **Translate the following paragraph into Chinese.**

网页可访问性不仅能完成一系列请求或者经验证是有效的，网页可访问性还关乎着高质量的设计。因此，最好能请一位屏幕读取和用户界面设计方面的专家测试你的网页，因为他们很熟悉网页的功能、用途以及浏览设计差的网站时会遇到的种种困难，并且能切实解决这些困难。

Text B

Ⅰ. **Read the passage above and decide whether the following statements are true (T) or false (F).**

1. T　2. F　3. F　4. F　5. T　6. F　7. F

Ⅱ. **Fill in the blanks with the information given in the text.**

1. bug-free and reliable

2. correctness; completeness; quality
3. tactful; the bugs
4. document; objectives; scope; approach; focus
5. designated testing activities; evaluation

Writing

1. Goals	2. Schedule	3. Cases	4. Code	5. Interface
6. Environment	7. Hardware	8. Tools	9. Bug	10. Release

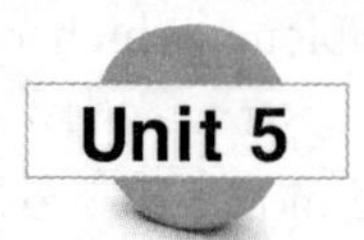

Unit 5

Text A

Ⅰ. **Answer the following questions briefly according to the text.**

1. Careful research, proper introduction and good follow-up.
2. Find out what groups your products or services will attract the most. Find companies that need your products or services in the area. Carefully research your target market for an assessment of the things you may be able to offer.
3. Help to find areas of demands your company can fill in.
4. Who you are, what you offer, where you are located, and how to contact you.
5. Grand opening date, special sales, top-selling products, number of customers to date, and company slogan/logo.
6. You can give a call back or a reply message. Offer coupons, discounts or a free item.

Ⅱ. **Match the words with their explanations in the right column.**

1. assessment	come to understand
2. respond	give particular importance to sth.
3. take off	get down to
4. figure out	a person that is competing with another
5. emphasize	a process to make a judgment about a person
6. start out	achieve sudden growth or success
7. competitor	reply

Ⅲ. **Complete the sentences with the words or phrases given in the table. Change the form if necessary.**

1. coupon	2. specialties	3. follow-up	4. logo	5. emphasizes
6. handout	7. assessment	8. poll	9. competitors	10. slogan

Ⅳ. Choose the best words or phrases to complete the following passage.

1—5 CDDBC 6—10 DCDCB 11—15 CDCBD

V. Translate the following paragraph into Chinese.

惠普个人计算机部门的领军者不辞辛劳地带领着惠普打败竞争对手戴尔和联想，成功地将惠普推上了全球个人计算机市场的第一把交椅，同时收获了丰厚的利润。布拉德利的战略是令人耳目一新的设计和引人瞩目的市场推广活动。说唱明星周杰伦和网球明星威廉姆斯在广告中说，用惠普计算机在 Windows 启动之前就能看电影，浏览照片，编辑音乐。电视广告以这句宣传口号结束：“计算机再一次回归个性”。

Text B

Ⅰ. Choose the best answers according to the passage.

1. C 2. B 3. A 4. C 5. D

Ⅱ. Translating.

A. Translate the following sentences into English.

1. They decide to move to the suburb of Shanghai next year.
2. I can't stop now—I'm a bit pressed for time.
3. Her shop now makes profits of over $1,000,000 a year.
4. Lily prepared a dress with matching accessories for the wedding.
5. The country has the capability to produce nuclear weapons.

B. Translate the following sentences into Chinese.

1. 我已经租了一个房子，并且付了租金。
2. 他在我对面坐下，对着我笑。
3. 尽管她 80 多岁了，仍然精力十足。
4. 吉姆不断地改变主意。
5. 她希望将来在电影业能有长远的发展。

Writing

1. Terms of payment are specified in great detail in our contract.
2. We are sure that this initial order will result in further business in the future.
3. As we are in badly demand of the goods, we shall appreciate it if you dispatch the goods at early date.
4. We'll ensure the prompt dispatch of the goods to your port.
5. As your quotation is competitive and acceptable to us, we are considering orders of large quantities.
6. We can't accept any fresh orders because we are fully committed. But as soon as fresh supplies come in, we shall contact you without delay.

Unit 6

Text A

Ⅰ. **Answer the following questions briefly according to the text.**

1. Consumer packaging aims at containing the goods, promoting the sale of it and facilitating use of it.
2. First, it should protect the goods from damage during handling, storing and transportation. Second, it should promote logistical efficiency. Third, it is the communication or information transfer.
3. Vibration, impact, puncture or compression.
4. The size, shape, and type of packaging material influence the type and amount of material handling equipment as well as how goods are stored in the warehouse. Likewise, package size and shape affect transportation in loading and unloading.
5. To identify package contents for receiving, order selection and shipment verification, etc.
6. The final communication role of logistics packaging is to provide instructions as to how to handle the cargo and how to prevent possible damage.

Ⅱ. **Fill in the blanks with the words or phrases given below. Change the form if necessary.**

1. package	2. essential	3. potential	4. consumers
5. information	6. motivate	7. purchase	8. distinctive
9. comply with	10. Protecting	11. transport	12. shipping

Ⅲ. **Match the indicative marks and the warning marks with their Chinese meanings.**

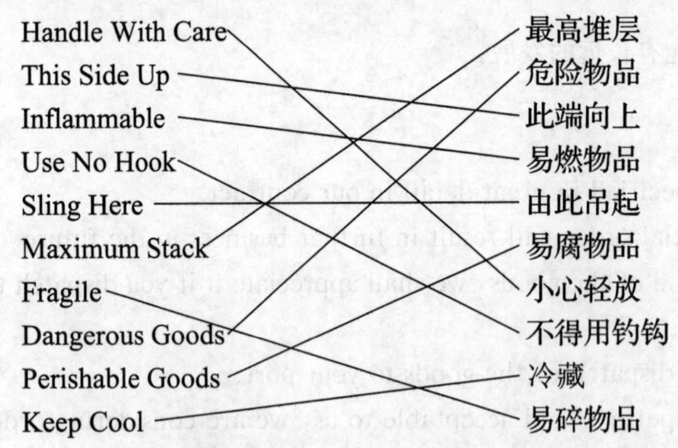

Ⅳ. **Read the text again and fill in the blanks with the information given in the text.**

1. major competitive force	2. industrial packaging; consumer packaging
3. productivity of logistics	4. handling; storing; transportation
5. the transportation rate	6. a great freedom of choice

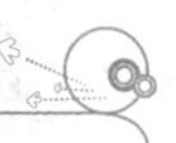

Ⅴ. Translate the following paragraph into Chinese.

包装是实现商品价值最重要的方法之一。它可以保护、美化商品，并且是构成商品储存、运输、销售的重要过程。包装标志是为了方便货物运输、装卸及储存，便于识别货物和防止货物损坏而在货物外包装上刷写的标志。包装标志主要包括运输标志、指示性标志、警告性标志等。

Text B

Ⅰ. Choose the best answers according to the passage.

1. D　2. C　3. D　4. A　5. B

Ⅱ. Fill in each blank with the appropriate form of the word given in the brackets.

1. ultimately　2. encountered　3. extensive　4. adequate
5. liability　6. endure　7. excess　8. sweating

Writing

1. the locked position　2. color photo　3. editing the image
4. protection switch　5. complete its self-test　6. preview window

Unit 7

Text A

Ⅰ. Answer the following questions briefly according to the text.

1. Product strategy, price strategy, promotion strategy, and channel strategy.
2. How to use the cross-regional platform to break down barriers of resources, optimize the resource combination and improve the brand value, has become a hot talk and concerned research of many enterprises.
3. The powerful features of the network contain almost all of the advantages of media advertising: the network multimedia acoustic-optic function and the three dimensional animation features in advertising planning, and the chat function of network.
4. Combine with the related industry company, and build sale spots to sell series of products on the Internet together. Using this kind of means can increase the consumer Internet willingness and consumption motivation.

Ⅱ. Match the words on the left with their meanings on the right.

1. D　2. H　3. G　4. A　5. F　6. E　7. C　8. B

Ⅲ. Complete the sentences with the words from Ⅱ. Change the form if necessary.

1. increment　2. dimensional　3. analyzed　4. optimize
5. Elasticity　6. characteristics　7. consumption　8. integrated

Text B

Ⅰ. **Read the passage above and decide whether the following statements are true (T) or false (F).**

1. T 2. F 3. T 4. F

Ⅱ. **Complete the following sentences by translating the Chinese in the brackets.**

1. potential for promotion   2. Training courses
3. company's management   4. references
5. sound scripted

Ⅲ. **Translate the following paragraph into Chinese.**

他建议这样说："我的应聘者在过去四年一直都是他们公司的顶尖销售人员，他在这一时期获得了两次升职机会，他是唯一快速被邀请到公司管理层的销售人员"。"这位应聘者的推荐信是我见过的最好的，他表示有兴趣去你们公司您所在的部门工作。所以我今天联系了您和另外三家公司谈关于这位应聘者的事。他下周已经有两个面试计划，所以……我什么时候能拟订出您的面试时间？"

Writing

China Nanjing Qiliang Imp. & Exp. Corp.
256 Ninghai Road 210096
Nanjing, China
Tel:(025) 83596388
Fax:(025) 83596387
E-Mail: njqiliang@vip. sina. com. cn

April 8th, 2003
Boston Electrics, Inc.
78 Quicy Rd.
Boston, MA 02127
U. S. A.

Gentlemen,

Having obtained from the Web that your business scope coincides with us, we are writing to you in the hope of establishing business relations with you.

We have been in this line for many years and now we are one of the largest importers of electric goods in Nanjing. As to our credit standing, we are permitted to mention the Bank of China, Nanjing, as a reference.

At present we are interested in your cordless phone and we look forward to hearing from you soon.

Yours truly,
The Fareast Trading Co. Ltd.

Unit 8

Text A

Ⅰ. Read the passage above and decide whether the following statements are true (T) or false (F).

1. F　　2. T　　3. F　　4. F　　5. T

Ⅱ. Fill in each blank with the appropriate form of the word given in brackets.

1. efficiency　　2. medium　　3. complicated　　4. aim at
5. be satisfied with　　6. constrain　　7. invoice　　8. Seek out

Ⅲ. Match the following terms with their Chinese meaning.

1. E　　2. D　　3. C　　4. A　　5. B　　6. K
7. L　　8. I　　9. G　　10. F　　11. J　　12. H

Ⅳ. Translating.

A. Translate the following passage into Chinese.

在电子商务环境中，人们利用万维网或因特网做生意，买卖双方的接触不再需要面对面，而是通过网络。在网上，买卖双方都有更多机会了解对方，不仅是一对一，而且可以多对多。无地域限制，无时间限制，双方都能找到称心如意的对家。

B. Translate the following sentences into English.

1. Managers can configure the devices from the browser.

2. I hope you'll be entirely satisfied with this initial shipment.

3. Can the government eliminate poverty?

4. The police used horses to constrain the crowds from approaching the conference hall.

5. Seek out readers' criticisms and opinions.

Text B

Ⅰ. Read the passage above and decide whether the following statements are true (T) or false (F).

1. T　　2. F　　3. F　　4. T　　5. F　　6. T

Ⅱ. Complete the following sentences by translating the Chinese in the brackets.

1. To get the eBay business off the ground

2. became a popular channel for auctioning collectibles
3. garage sales; collectibles shows; flea markets; classified advertisements
4. a nonrefundable insertion fee; a final value fee
5. notifies the buyer and the seller via e-mail

Writing

Dec. 15

Dear Jenny,

I have two tickets for a famous Christmas concert at Beijing Concert Hall on Friday, December 25. I heard it was a wonderful concert and I made great effort to get the ticket. Will you join me? I'll be waiting for you at eight sharp Friday night in front of the concert hall, so don't disappoint me!

Warmest regards,

Alice

Unit 9

Text A

Ⅰ. **Answer the following questions briefly according to the text.**

1. Since EB is conducted on Internet, and the Internet is such an open system.
2. The data may be stolen, corrupted, misused altered or falsely generated. Computer system or website may be attacked and make render systems unable to operate properly, hardware or software is used illegally.
3. The heavy lessons of Internet perpetration draw people's attention to improving the situation of Internet and EB security.
4. That means all the electronic data are protected from unauthorized disclosure.
5. The EB security requirement is that the content and privacy of electronic transaction must be protected from being intercepted, abused, altered, deleted or disturbed during the electronic data interchange processing, transmitting and storing.

Ⅱ. **Fill in each blank with the appropriate form of the word given in bracket.**

1. associated with　2. generation　3. transmission　4. illegally

5. disclosure　　6. adequate　　7. integrity

Ⅲ. **Fill in the table below by giving the corresponding Chinese or English equivalents.**

exponential growth	爆炸式增长
head-on blow	当头棒喝
EB applications	电子商务应用
electronic transaction	电子交易
unauthorized disclosure	非授权暴露
confidential data	机密数据
firewall	防火墙
security assurance system	安全确认系统
digital signature	数字签名
digital certificate	数字证书

Ⅳ. **Translating.**

A. Translate the following sentences into Chinese.

1. 在网络给人们无尽方便与机会的同时，它也给人们带来了风险。
2. 它发展非常快，前景诱人，多功能服务和在网上设立的爆炸式的增长速度吸引越来越多的公司和个人像淘金热般地加入进来。
3. 另一个原因是人们如此急切地投入网络和电子商务应用，以至于他们来不及考虑可能涉及的细节。

B. Translate the following sentences into English.

1. Associate with positive people and protect yourself from all types of negativity.
2. As I am eager to be a teacher in the future, it did arouse my curiosity to do this research.
3. Saturday's accident is a head-on blow to our plan.
4. In response to these questions, we can take some effective methods to resolve it.
5. This project will bring profit, and it also will bring risk at the same time.

Text B

Ⅰ. **Read the passage above and decide whether the following statements are true (T) or false (F).**

1. T　　2. T　　3. F　　4. F　　5. T

Ⅱ. **Translate the following paragraph into Chinese.**

1. 网络安全，是计算机信息系统安全的一个重要方面。如同打开了的潘多拉魔盒，计算机系统的互联，在大大扩展信息资源的共享空间的同时，也将其本身暴露在更多恶意攻击之下。

2. 采用目前最先进的漏洞扫描系统定期对工作站、服务器、交换机等进行安全检查，并根据检查结果向系统管理员提供详细可靠的安全性分析报告，为提高网络安全整体水平产生重要依据。

3. 总地来说，网络安全是一个综合性问题。它涉及技术、管理、使用和许多其他方面，包括它自己的信息系统安全问题，有物理和逻辑的技术措施。作为一种技术，它只能解决问题的一方面，而不是万能的。

Writing

A regular monthly meeting of Nortel Marketing is going to be held at Conference Room 800，Sky Hotel，9:00 am to 11:00 am，September 14，2012. Three managers（Mr. John，Mr. Green and Mrs. Wang）in this area will make their reports at the meeting. Write an agenda for the meeting by referring to the format given above.

Date：	14th，September，2012
Time：	9:00 am to 11:00 am
Venue：	Conference Room800，Sky Hotel
Participants：	Mr. John，Mr. Green and Mrs. Wang
Reading of minutes of previous meeting and approval of the minutes	
1. Month Marketing Analysis 2. How to Manage Marketing 3. Marketing Prospect	by Mr. John by Mr. Green by Mrs. Wang
Discussion on the unfinished business（mentioned at the previous meeting）：	
Discussion of the new business（presented in the reports）：	
Date of next meeting：	
Other business：	
Adjournment（休会）：	

Text A

Ⅰ. **Fill in the blanks with the information given in the text.**

1. used；replace　2. combined；cause　3. affecting　4. resulting in

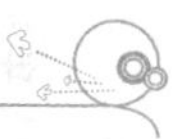

Ⅱ. **Match each of the following terms to its equivalent(s).**

1. F　　2. A　　3. E　　4. C　　5. D　　6. B

Ⅲ. **Read the passage above and decide whether the following statements are true (T) or false (F).**

1. F　　2. F　　3. T　　4. T

Ⅳ. **Translate the following passage from English into Chinese.**

即时消息就像与另一个人或一群人实时谈话。当你输入和发送一条即时消息时，所有参与者马上会看到这条消息。与电子邮件不同，所有的参与者都必须在线（或者联网），并且同时坐在他们各自的计算机前。通过即时消息交流称作聊天。

Text B

Ⅰ. **Read the passage above and decide whether the following statements are true (T) or false (F).**

1. T　　2. T　　3. F

Ⅱ. **Translate the following sentences into Chinese**

1. 黑客设计的病毒程序会像疾病一样在计算机间传播。一旦计算机感染了病毒，它会删除文件，私发邮件、甚至把私人信息泄露给犯罪分子。
2. 木马则是最阴险狡诈的病毒。像传说中的特洛伊木马，它也善用诡计，一不小心就可能从网上下载木马。木马病毒会伪装成游戏或者常用软件，有些木马还会给计算机开新的“后门”，让“犯罪分子”轻易侵入你的计算机获取信息。

Writing

Dear girls and boys in Class 0403,

This Saturday evening, from 7 pm to 9 pm, at Singeing dining hall, there will be a ball held by Class 0301. We warmly invite you to take part in it. There you can enjoy yourselves with dance, music and various games. And we'll have a chance of communication, which will enhance the friendship between us. Come and enjoy ourselves together.

Yours,

Mike

Monitor of Class 0301

Unit 11

Text A

Ⅰ. **Read the passage and answer the following questions.**

1. 21-day period.
2. 5 to 7 workdays.
3. The customer has to call Technical Support before exchanging a product .
4. The return material authorization number.
5. The company will pay for the return shipping.

Ⅱ. **Read the passage above and decide whether the following statements are true (T) or false (F).**

1. F　　2. T　　3. F　　4. T

Ⅲ. **Put the following steps of returning or exchanging into correct order. Write numbers before the statements.**

6-4-1-3-7-2-5

Ⅳ. **Fill in each blank with the appropriate form of the word given in brackets.**

1. exchange	2. package	3. replace	4. authorization
5. subtract	6. original	7. receipt	8. bold

Ⅴ. **Translating.**

A. Translate the following sentences into Chinese.

1. 我们承担退货运输费用。
2. 您可以在订单状态中查看您退换货的进度。
3. 对于换货，我们需要在收到您寄回的货品后才能将新货品发出。
4. 不遵守此流程可能延误您信用额度或者换货进度，同时您需承担导致损失的风险以及寄回产品所需的费用。

B. Translate the following sentences into English.

1. Users should not install any software in any workstation without authorization.
2. We undertake to replace the specifications.
3. They give 10% discount for cash payment.
4. It is very convenient to pay by credit card.
5. We'll refund your money if you aren't satisfied.

Text B

Ⅰ. **Read the passage above and decide whether the following statements are true (T) or false (F).**

1. T 2. F 3. F 4. T 5. T

Ⅱ. **Choose from the following words and expressions and fill in the appropriate box.**

1. specific 2. defective 3. confusion 4. lashes out
5. mad 6. registered 7. refund 8. take advantage of

Writing

For Over-Shipment

Dear Sirs,

We thank you for promptness in delivering the Chinaware we ordered on 20th Dec. The number of cartons delivered by your carrier this morning was 360, whereas our order was for only 320.

Unfortunately, our present needs are completely covered and we cannot make use of the extra goods. Please inform us by fax what we are to do with the extra goods.

Yours faithfully,

Unit 12

Text A

Ⅰ. **Fill in each blank with the appropriate form of the words given in brackets.**

1. motivation 2. wireless 3. prevalent 4. embed
5. collaboration 6. diagnostic 7. psychological 8. auditory

Ⅱ. **Translating.**

A. Translate the following sentences into Chinese.

1. 由教师本人指导一群学生的传统教学方式依然盛行，但学校发展得越来越依靠软件教学。

2. 技术能使非音乐家创造出音乐，如利用人脑电波作曲的声控作曲系统和软件。

B. Translate the following sentences into English with the given words.

1. In the future the ability to embed a video the newspapers that a real

multimedia newspaper.

2. She wrote the book in collaboration with one of her students.

3. The motivation for the decision is the desire to improve our service to our customers.

Text B

Ⅰ. **Red the passage above and decide whether the following statements are true (T) or false (F)..**

1. F 2. T 3. F 4. T 5. F 6. T

Ⅱ. **Complete the following sentences by translating the Chinese in the brackets.**

1. provide the infrastructure/platform to deliver videos
2. The distribution of video content
3. amateur productions dominate
4. organization marketing to consumers
5. attempting to generate revenue through

Writing

1. PERSONAL INFORMATION
2. OBJECTIVE
3. EDUCATION
4. WORK EXPERIENCE
5. SKILLS
6. PERSONALITY
7. INTEREST

References

[1] 陈建峡. 计算机英语 [M]. 武汉：华中科技大学出版社，2006.
[2] 冯国华. 计算机英语 [M]. 北京：机械工业出版社，2012.
[3] 教育部《计算机英语》教材编写组. 计算机英语 [M]. 北京：高等教育出版社，2001.
[4] 任军战. 计算机英语 [M]. 北京：外语教学与研究出版社，2008.
[5] 谭新星，段琢华. IT 行业英语 [M]. 广州：暨南大学出版社，2012.
[6] 伍忠杰. IT 英语 [M]. 上海：上海外语教育出版社，2009.
[7] 徐小贞，等. 新职业英语　IT 英语形成性评估手册 [M]. 北京：外语教学与研究出版社，2009.
[8] 张玲，等. 计算机英语 [M]. 北京：清华大学出版社，2008.